Charles Louis Flint

Milch cows and dairy farming

Comprising the breeds, breeding, and management in health and disease

Charles Louis Flint

Milch cows and dairy farming
Comprising the breeds, breeding, and management in health and disease

ISBN/EAN: 9783337145651

Printed in Europe, USA, Canada, Australia, Japan

Cover: Foto ©berggeist007 / pixelio.de

More available books at **www.hansebooks.com**

MILCH COWS

AND

DAIRY FARMING.

COMPRISING

THE BREEDS, BREEDING, AND MANAGEMENT IN HEALTH AND
DISEASE, OF DAIRY AND OTHER STOCK; THE SELECTION
OF MILCH COWS, WITH A FULL EXPLANATION OF
GUENON'S METHOD; THE CULTURE OF
FORAGE PLANTS, &C.

BY

CHARLES I FLINT,

AUTHOR OF "GRASSES AND FORAGE PLANTS," &C., &C.

BOSTON:
J. E. TILTON AND COMPANY.
1867.

To

THE MASS. STATE BOARD OF AGRICULTURE,

THE

MASS. SOCIETY FOR THE PROMOTION OF AGRICULTURE,

AND THE VARIOUS

AGRICULTURAL SOCIETIES OF THE UNITED STATES,

WHOSE EFFORTS HAVE CONTRIBUTED SO LARGELY TO IMPROVE THE

DAIRY STOCK OF OUR COUNTRY

This Treatise,

DESIGNED TO ADVANCE THAT HIGHLY IMPORTANT INTEREST,

IS RESPECTFULLY DEDICATED,

BY

THE AUTHOR.

PREFACE.

THIS work is designed to embody the most recent information on the subject of dairy farming. My aim has been to make a practically useful book. With this view, I have treated of the several breeds of stock, the diseases to which they are subject, the established principles of breeding, the feeding and management of milch cows, the raising of calves intended for the dairy, and the culture of grasses and plants to be used as fodder.

For the chapter on the diseases of stock, I am largely indebted to Dr. C. M. Wood, Professor of the Theory and Practice of Veterinary Medicine, and to Dr. Geo. H. Dadd, Professor of Anatomy and Physiology, both of the Boston Veterinary Institute. If this chapter contributes anything to promote a more humane and judicious treatment of cattle when suffering from disease, I shall feel amply repaid for the labor bestowed upon the whole work.

The chapter on the Dutch dairy, which I have translated from the German, will be found to be of great practical value, as suggesting much that is applicable to our American dairies. This chapter has never before, to my knowledge, appeared in English.

The full and complete explanation of Guénon's method of judging and selecting milch cows, — a method originally regarded as theoretical, but now generally admitted to be very useful in practice, — I have translated from the last edition of the treatise of M. Magne, a very sensible French writer, who has done good service to the agricultural public by the clearness and simplicity with which he has 'freed' that system from its complicated details.

The work will be found to contain an account of the most enlightened practice in this country, in the statements of those actually engaged in dairy farming; the details of the dairy husbandry of Holland, where this branch of industry is made a specialty to greater extent, and is consequently carried to a higher degree of perfection, than in any other part of the world; and the most recent and productive modes of management in English dairy farming, embracing a large amount of practical and scientific information, not hitherto presented to the American public in an available form.

Nothing need be said of the usefulness of a treatise on the dairy. The number of milch cows in the country, forming so large a part of our material wealth, and serving as a basis for the future increase and improvement of every class of·neat stock, on which the prosperity of our agriculture mainly depends; the intrinsic value of milk as an article of internal commerce, and as a most healthy and nutritious food; the vast quantity of it made into butter and cheese, and used in every family; the endless details of the management, feeding, and treatment, of dairy stock, and the care and attention requisite to obtain from this branch of farming the highest profit, all concur to make the want of such a treatise, adapted to our climate and circumstances, felt not only by practical farmers, but· by a large class of consumers, who can appreciate every improvement which may be made in preparing the products of the dairy for their use.

The writer has had some years of practical experience in the care of a cheese and butter dairy, to which has been added a wide range of observation in some of the best dairy districts of the country; and it is hoped that the work now submitted to the public will meet that degree of favor usually accorded to an earnest effort to do something to advance the cause of agriculture.

DAIRY FARMING.

CHAPTER I.

INTRODUCTORY.—THE VARIOUS RACES OF PURE
BRED CATTLE IN THE UNITED STATES.

THE milking qualities of our domestic cows are, to some extent, artificial, the result of care and breeding. In the natural or wild state, the cow yields only enough to nourish her offspring for a few weeks, and then goes dry for several months, or during the greater part of the year. There is, therefore, a constant tendency to revert to that condition, which is prevented only by judicious treatment, designed to develop and increase the milking qualities so valuable to the human race. If this judicious treatment is continued through several generations of the same family or race of animals, the qualities which it is calculated to develop become more or less fixed, and capable of transmission. Instead of being exceptional, or peculiar to an individual, they become the permanent characteristics of a breed. Hence the origin of a great variety of breeds or races, the characteristics of each being due to local circumstances, such as climate, soil, and the special objects of the breeder, which may be the production of milk, butter and cheese, or the raising of beef or working cattle.

A knowledge of the history of different breeds, and

especially of the dairy breeds, is of manifest importance. Though very excellent milkers will sometimes
be found in all of them, and of a great variety of forms,
the most desirable dairy qualities will generally be
found to have become fixed and permanent characteristics of some to a greater extent than of others ; but
it does not follow that a race whose milking qualities
have not been developed is of less value for other purposes, and for qualities which have been brought out
with greater care. A brief sketch of the principal
breeds of American cattle, as well as of the grades or
the common stock of the country, will aid the farmer,
perhaps, in making an intelligent selection with reference to the special object of pursuit, whether it be the
dairy, the production of beef, or the raising of cattle for
work.

In a subsequent chapter on the selection of milch
cows, the standard of perfection will be discussed in
detail, and the characteristics of each of the races will
naturally be measured by that. In this connection, and
as preliminary to the following sketches, it may be
stated that, whatever breed may be selected, a full supply of food and proper shelter are absolutely essential
to the maintenance of any milking stock, the food of
which goes to supply not only the ordinary waste of
the system common to all animals, but also the milk
secretions, which are greater in some than in others.
A large animal on a poor pasture has to travel much
further to fill itself than a small one. A small or
medium-sized cow would return more milk in proportion to the food consumed, under such circumstances,
than a large one.

In selecting any breed, therefore, regard should be
had to the circumstances of the farmer, and the object
to be pursued. The cow most profitable for the milk-

dairy may be very unprofitable in the butter and cheese dairy, as well as for the production of beef; while for either of the latter objects the cow which gave the largest quantity of milk might prove very unprofitable. It is desirable to secure a union and harmony of all good qualities, so far as possible ; and the farmer wants a cow that will milk well for some years, and then, when dry, fatten readily, and sell to the butcher for the highest price. These qualities, though often supposed to be incompatible, will be found to be united in some breeds to a greater extent than in others; while some peculiarities of form have been found, by observation, to be better adapted to the production of milk and beef than others. This will appear in the following pages.

Fig. 1. Ayrshire Cow, imported and owned by Dr. Geo. B. Loring, Salem, Mass.

THE AYRSHIRES are justly celebrated throughout Great Britain and this country for their excellent dairy qualities. Though the most recent in their origin, they are pretty distinct from the other Scotch and English races. In color, the pure Ayrshires are generally red and white, spotted or mottled, not roan like many of the short-horns, but often presenting a bright contrast of colors. They are sometimes, though rarely, nearly or quite all red, and sometimes black and white ; but the favorite color is red and white brightly contrasted, and by some, strawberry-color is preferred. The head is small, fine, and clean ; the face long, and narrow at the muzzle, with a sprightly yet generally mild expression ; eye small, smart, and lively ; the horns short, fine, and slightly twisted upwards, set wide apart at the roots ; the neck thin ; body enlarging from fore to hind quarters ; the back straight and narrow, but broad across the loin ; joints rather loose and open ; ribs rather flat ; hind quarters rather thin ; bone fine ; tail long, fine and bushy at the end ; hair generally thin and soft ; udder light color and capacious, extending well forward under the belly ; teats of the cow of medium size, generally set regularly and wide apart ; milk-veins prominent and well developed. The carcass of the pure-bred Ayrshire is light, particularly the fore quarters, which is considered by good judges as an index of great milking qualities ; but the pelvis is capacious and wide over the hips.

On the whole, the Ayrshire is good-looking, but wants some of the symmetry and aptitude to fatten which characterize the short-horn, which is supposed to have contributed to build up this valuable breed on the basis of the original stock of the county of Ayr ; a county extending along the eastern shore of the Frith of Clyde, in the south-western part of Scotland, and divided into three districts, known as Carrick, Cunningham, and Kyle : the first famous as the lordship of Robert Bruce, the last for the production of this, one of the most remarkable dairy breeds of cows in the world. The original stock of this county, which undoubtedly formed the basis of the

present Ayrshire breed, are described by Aiton, in his
Treatise on the Dairy Breed of Cows, as of a diminutive size, ill fed, ill shaped, and yielding but a scanty
return in milk. They were mostly of a black color,
with large stripes of white along the chine and ridge of
their backs, about the flanks, and on their faces. Their
horns were high and crooked, having deep ringlets at
the root,—the plainest proof that the cattle were but
scantily fed; the chine of their backs stood up high and
narrow; their sides were lank, short, and thin; their
hides thick, and adhering to their bones; their pile was
coarse and open; and few of them yielded more than
six or eight quarts of milk a day when in their best
plight, or weighed when fat more than from twelve to
sixteen or twenty stones avoirdupois, at eight pounds
the stone, sinking offal.

" It was impossible," he continues, " that these cattle,
fed as they then were, could be of great weight, well
shaped, or yield much milk. Their only food in winter
and spring was oat-straw, and what they could pick up
in the fields, to which they were turned out almost
every day, with a mash of weak corn and chaff daily for
a few days after calving; and their pasture in summer
was of the very worst quality, and eaten so bare that
the cattle were half starved, and had the aspect of
starvelings. A wonderful change has since been made
in the condition, aspect, and qualities, of the Ayrshire
dairy stock. They are not now the meagre, unshapely
animals they were about forty years ago; but have
completely changed into something as different from
what they were then as any two breeds in the island
can be from each other. They are almost double the
size, and yield about four times the quantity of milk
that the Ayrshire cows then yielded. They were not
of any specific breed, nor uniformity of shapes or color:

2

neither was there any fixed standard by which they could be judged."

Aiton wrote in 1815, and even then the Ayrshire cattle had been completely changed from what they were in 1770, and had, to a considerable extent, at least, settled down into a breed with fixed characteristics, distinguished especially for an abundant flow and a rich quality of milk. A large part of the improvement then manifested was due to better feeding and care, but much, no doubt, to judicious crossing. Strange as it may seem, considering the modern origin of this breed, "all that is certainly known is that a century ago there was no such breed as Cunningham or Ayrshire in Scotland. Did the Ayrshire cattle arise entirely from a careful selection of the best native breed ? If they did, it is a circumstance unparalleled in the history of agriculture. The native breed may be ameliorated by careful selection ; its value may be incalculably increased ; some good qualities, some of its best qualities, may be for the first time developed; but yet there will be some resemblance to the original stock, and the more we examine the animal the more clearly we can trace out the characteristic points of the ancestor, although every one of them is improved."

Aiton remembered well the time when some short-horn or Dutch cattle, as they were then called, were procured by some gentlemen in Scotland, and particularly by one John Dunlop, of Cunningham, who brought some Dutch cows — doubtless short-horns — to his byres soon after the year 1760. As they were then provided with the best of pasture, and the dairy was the chief object of the neighborhood, these cattle soon excited attention, and the small farmers began to raise up crosses from them. This was in Cunningham, one of the districts of Ayrshire, and Mr. Dunlop's were.

without doubt, among the first of the stranger breed
that reached that region. About 1750, a little previous
to the above date, the Earl of Marchmont bought of the
Bishop of Durham several cows and a bull of the Teeswater breed, all of a brown color spotted with white,
and kept them some time at his seat in Berwickshire.
His lordship had extensive estates in Kyle, another district of Ayrshire, and thither his factor, Bruce Camp-
bell, took some of the Teeswater breed and kept them
for some time, and their progeny spread over various
parts of Ayrshire. A bull, after serving many cows of
the estates already mentioned, was sold to a Mr. Hamil-
ton, in another quarter of Ayrshire, and raised a numerous offspring.

About the year 1767, also, John Orr sent from Glasgow to his estate in Ayrshire some fine milch cows, of
a much larger size than any then in that region. One
of them cost six pounds, which was more than twice
the price of the best cow in that quarter. These cows
were well fed, and of course yielded a large return of
milk; and the farmers, for miles around, were eager to
get their calves to raise.

About the same time, also, a few other noblemen and
gentlemen, stimulated by example, bought cattle of the
same appearance, in color brown spotted with white, all
of them larger than the native cattle of the county,
and when well fed yielding much larger quantities of
milk, and their calves were all raised. Bulls of their
l reed and color were preferred to all others.

From the description given of these cattle, there is
no doubt that they were the old Teeswater, or Dutch;
the foundation, also, according to the best authorities, of
the modern improved short-horns. With them and the
crosses obtained from them the whole county gradually became stocked; and supplied the neighboring

counties, by degrees, till at present the whole region, comprising the counties of Ayr, Renfrew, Lanark, Dumbarton, and Stirling, and more than a fourth part of the whole population of Scotland, a large proportion of which is engaged in manufactures and commercial or mechanical pursuits, furnishing a ready market for milk and butter, is almost exclusively stocked with Ayrshires.

The cross with larger cattle and the natives of Ayr- shire produced, for many years, an ugly-looking beast, and the farmers were long in finding out that they had violated one of the plain principles of breeding in coupling a large and small breed so indiscriminately together, especially in the use of bulls proportionately larger than the cows to which they were put. They did not then understand that no crosses could be made in that way to increase the size of a race, without a corresponding increase in the feed; and many very ill-shaped animals were the consequence of ignorance of a natural law. They made large bones, but they were never strong and vigorous in proportion to their size. Trying to keep large animals on poor pasture produced the same effect. The results of first crosses were therefore very unsatisfactory; but gradually bet- ter feeding and a reduction in size came to their aid, while in the course of years more enlightened views of farming led to higher cultivation, and consequently to higher and better care and attention to stock. The effect of crosses with the larger Teeswater or short- horn was not so disastrous in Ayrshire as in some of the mountain breeds, whose feed was far less, while their exposure on high and short pastures was greater.

The climate of Ayrshire is moist and mild, and the soil rich, clayey, and well adapted to pasturage, but difficult to till. The cattle are naturally hardy and active, and capable of enduring severe winters, and

of easily regaining condition with the return of spring
and good feed. The pasture-land of the county is
devoted to dairy stock, — chiefly for making butter
and cheese, a small part only being used for fattening
cows when too old to keep for the dairy. The breed
has undergone very marked improvements since Aiton
wrote, in 1815. The local demand for fresh dairy products has very naturally taxed the skill and judgment of
the farmers and dairy-men to the utmost, through a
long course of years ; and thus the remarkable milking
qualities of the Ayrshires have been developed to such a
degree that they may be said to produce a larger quantity of rich milk and butter in proportion to the food
consumed, or the cost of production, than any other of
the pure-bred races. The owners of dairies in the
county of Ayr and the neighborhood were generally
small tenants, who took charge of their stock themselves, saving and breeding from the offspring of good
milkers, and drying off and feeding such as were found
to be unprofitable for milk, for the butcher ; and thus the
production of milk and butter has for many years been
the leading object with the owners of this breed, and
symmetry of form and perfection of points for any other
object have been very much disregarded, or, if regarded
at all, only from this one point of view — the production of the greatest quantity of rich milk.

The manner in which this result has been brought
about may further be seen in a remark of Aiton, who
says that the Ayrshire farmers prefer their dairy bulls
according to the feminine aspect of their heads and
necks, and wish them not round behind, but broad at
the hook-bones and hips, and full in the flanks. This
was more than forty years ago, and under such circumstances, and with such care in the selection of bulls and
cows with reference to one specific object, it is not

surprising that we find a breed now wholly unsurpassed when the quantity and quality of their produce is con- sidered with reference to their proportional size and

Fig. 2.   Ayrshire Bull "ALBERT,"
Imported and owned by the Mass. Soc. for Promoting Agriculture.

the food they consume.   The Ayrshire cow has been known to produce over ten imperial gallons of good milk a day.

A cow-feeder in Glasgow, selling fresh milk, is said to have realized two hundred and fifty dollars in seven months from one good cow; and it is stated, on high authority, that a dollar a day for six months of the year is no uncommon income from good cows under similar circumstances, and that seventy-five cents a day is be- low the average.   But this implies high and judicious feeding, of course : the average yield, on ordinary feed would be considerably less.

Youatt estimates the daily yield of an Ayrshire cow, for the first two or three months after calving, at five gallons a day, on an average ; for the next three months, at three gallons; and for the next four months, at one gallon and a half.   This would be 850 gallons as the

annual average of a cow; but, allowing for some unpro-
ductive cows, he estimates the average of a dairy at 600
gallons per annum for each cow. Three gallons and a
half of the Ayrshire cow's milk will yield one and a
half pounds of butter. He therefore reckons 257
pounds of butter, or 514 pounds of cheese, at the rate
of 24 pounds to 28 gallons of milk, as the yield of every
cow, at a fair and perhaps rather .low average, in an
Ayrshire dairy, during the year. Aiton sets the yield
much higher, saying that "thousands of the best Ayrshire dairy-cows, when in prime condition and well fed,
produce 1000 gallons of milk per annum; that in general three and three quarters to four gallons of their milk
will yield a pound and a half of butter; and that 27½
gallons of their milk will make 21 pounds of full-milk
cheese." Mr. Rankin puts it lower—at about 650 to
700 gallons to each cow; on his own farm of inferior
soil, his dairy produced an average of 550 gallons only.

One of the four cows originally imported into this
country by John P. Cushing, Esq., of Massachusetts,
gave in one year 3864 quarts, beer measure, or about
966 gallons, at ten pounds to the gallon, being an average of over ten and a half beer quarts a day for the
whole year. It is asserted, on good authority, that the
first Ayrshire cow imported by the Massachusetts Society for the Promotion of Agriculture, in 1837, yielded
sixteen pounds of butter a week, for several weeks in
succession, on grass feed only. These yields are not
so large as those stated by Aiton; but it should, perhaps, be recollected that our climate is less favorable
to the production of milk than that of England and Scotland, and that no cow imported after arriving at maturity could be expected to yield as much, under the same
circumstances, as one bred on the spot where the trial
is made, and perfectly acclimated.

In a series of experiments on the Earl of Chester-
field's dairy farm, at Bradley Hall, interesting as giving
positive data on which to form a judgment as to the
yield, it was found that, in the height of the season, the
Holderness cows gave 7 gallons 1 quart per diem; the
long-horns and Alderneys, 4 gallons 3 quarts; the Devons, 4 gallons 1 quart; and that, when made into butter,
the above quantities gave, respectively, 38½ ounces, 28
ounces, and 25 ounces.

The Ayrshire, a cow far smaller than the Holderness,
at 5 gallons of milk and 34 ounces of butter per day,
gives a fair average as to yield of milk, and an enormous production of butter, giving within 4½ ounces as
much from her 5 gallons as the Holderness from her 7
gallons 1 quart; her rate being nearly 7 ounces to the
gallon, while that of the Holderness is considerably
under 6 ounces.

The evidence of a large and practical dairyman is certainly of the highest value; and in this connection it
may be stated that Mr. Harley, the author of the *Harle-
ian Dairy System*, who established the celebrated Wil-
lowbank Dairy, in Glasgow, and who kept, at times,
from two hundred and sixty to three hundred cows,
always using the utmost care in selection, says that he
had cows, by way of experiment, from different parts
of the united kingdom. He purchased ten at one Edinburgh market, of the large short-horned breed, at
twenty pounds each, but these did not give more milk,
nor better in quality, than Ayrshire cows that were
bought at the same period for thirteen pounds a head;
and, on comparison, it was found that the latter were
much cheaper kept, and that they improved much more
in beef and fat in proportion to their size, than the
high-priced cows. A decided preference was therefore
given to the improved Ayrshire breed, from seven to

ten years old, and from eight to twenty pounds a head. Prime young cows were too high-priced for stall feeding; old cows were generally the most profitable in the long run, especially if they were not previously in good keeping. The cows were generally bought when near calving, which prevented the barbarous practice called hafting, or allowing the milk to remain upon the cow for a considerable time before she is brought to the market. This base and cruel custom is always pernicious to the cow, and in consequence of it she seldom recovers her milk for the season. The middling and large sizes of cows were preferred, such as weighed from thirty-five to fifty stone, or from five hundred to eight hundred pounds.

According to Mr. Harley, the most approved shape and marks of a good dairy cow are as follows: Head small, long, and narrow towards the muzzle; horns small, clear, bent, and placed at considerable distance from each other; eyes not large, but brisk and lively; neck slender and long, tapering towards the head, with a little loose skin below; shoulders and fore quarters light and thin; hind quarters large and broad; back straight, and joints slack and open; carcass deep in the rib; tail small and long, reaching to the heels; legs small and short, with firm joints; udder square, but a little oblong, stretching forward, thin-skinned and capacious, but not low hung; teats or paps small, pointing outwards, and at a considerable distance from each other; milk-veins capacious and prominent; skin loose, thin, and soft like a glove; hair short, soft, and woolly; general figure, when in flesh, handsome and well proportioned.

If this description of the Ayrshire cow be correct, it will be seen that her head and neck are remarkably clean and fine, the latter swelling gradually towards the

shoulders, both parts being unincumbered with superfluous flesh. The same general form extends backwards,
the fore quarters being light, the shoulders thin, and the
carcass swelling out towards the hind quarters, so that
standing in front of her it has the form of a blunted
wedge. Such a structure indicates very fully developed digestive organs, which exert a powerful influence
on the exercise of all the functions of the body, and
especially on the secretion of the milky glands, accompanied with milk-veins and udder partaking of the same
character as the stomach and viscera, being large and
capacious, while the external skin and interior walls of
the milk-glands are thin and elastic, and all parts
arranged in a manner especially calculated for the production of milk.

A cow with these marks will generally be of a quiet
and docile temper, which greatly enhances her value.
A cow that is of a quiet and contented disposition feeds
at ease, is milked with ease, and yields more than one
of an opposite temperament; while after she is past her
usefulness as a milker she will easily take on fat, and
make fine beef and a good quantity of tallow, because
she feeds freely, and when dry the food which went to
make milk is converted into fat and flesh. But there is no
breed of cows with which gentleness of treatment is so
indispensable as with the Ayrshire, on account of her
naturally nervous temperament. If she receive other
than kind and gentle treatment, she will often resent it
with angry looks and gestures, and withhold her milk;
and if such treatment is long continued, will dry up;
but she willingly and easily yields it to the hand that
fondles her, and all her looks and movements towards
her friends are quiet and mild.

As already remarked, the Ayrshires in their native
country are generally bred for the dairy, and no other

object; and the cows have obtained a just and world-wide reputation for this quality. The oxen are, however, very fair as working cattle, though they cannot be said to excel other breeds in this respect. The Ayrshire steer may be fed and turned at three years old, but for feeding purposes the Ayrshires are greatly improved by a cross with the short-horns, provided regard is had to the size of the animals. It is the opinion of good breeders that a high-bred short-horn bull and a large-sized Ayrshire cow will produce a calf which will come to maturity earlier, and attain greater weight, and sell for more money, than a pure-bred Ayrshire. This cross, with feeding from the start, may be sold fat at two or three years old, the improvement being especially seen in the earlier maturity and the size. Even Youatt, who maintains that the fattening properties of the Ayr-shires have been somewhat exaggerated, admits that they will fatten kindly and profitably, and that their meat will be good; while he also asserts that they unite, perhaps, to a greater degree than any other breed, the supposed incompatible qualities of yielding a great deal of milk and beef.

In the cross with the short-horn, the form becomes ordinarily more symmetrical, while there is, perhaps, little risk of lessening the milking qualities of the off-spring, if sufficient regard is paid to the selection of the individual animals to breed from. It is thought by some that in the breeding of animals it is the male which gives the external form, or the bony and muscu-lar system of the young, while the female imparts the respiratory organs, the circulation of the blood, the mucous membranes, the organs of secretion, &c.

If this principle is true, it follows that the milking qualities come chiefly from the mother, and that the bull can not materially alter the conditions which determine the transmission of these qualities, especially when they are as strongly marked as they are in the Ayrshire or the Jersey races. Others, however, maintain that it is more important to the perfection of their dairy to make a good choice of bulls than of heifers, because the property of giving much milk is more surely transmitted by the male than the female. Others still maintain that both parents are represented in the offspring, but that it is impossible to say beforehand what parts of the derivative system are to be ascribed to the one parent and what to the other, and that there is a blending and interfusion of the qualities of both which prevent the body of their progeny being mapped out into distinct regions, or divided into separate sets of organs, of which we can say, "This is from the father, that from the mother."

Till this question is settled, it is safe, in breeding for the dairy, to adhere to the rule of selecting only animals whose progenitors on both sides have been distinguished for their milking qualities. But where the history of either is unknown, a resort to a well-known breed, remarkable for its dairy qualities, is of no small importance; since, though the immediate ancestors of a male may not be known, if he belongs to a dairy breed, it is fair to presume that his progenitors were milkers. A study and comparison of the size and form of the milk mirror, and other points, indicated by Guénon, on a subsequent page, are worthy of careful consideration in selecting animals to breed from for the dairy, not only among pure-bred animals, but especially in crossing. In the scale of points adopted in England and this country as the standard of perfection for an Ayrshire cow, the udder, on which Guénon placed so much reliance, is valued at twelve times as much as that of the Devon, "because," as the judges affirm, "the Ayrshires have been bred almost exclusively with reference to their milking properties."

We must conclude, then, that "for purely dairy purposes the Ayrshire cow deserves the first place. In consequence of her small, symmetrical, and compact body, combined with a well-formed chest and a capacious stomach, there is little waste, comparatively speaking, through the respiratory system; while, at the same time, there is very complete assimilation of the food, and thus she converts a large proportion of her food into milk. So remarkable is this fact, that all dairy farmers who have any experience on the point agree in stating *that an Ayrshire cow generally gives a larger return of milk for the food consumed than a cow of any other breed.* The absolute quantity may not be so great, but it is obtained at a less cost; and this is the point upon which the question of profit depends."

I have dwelt thus at length upon this race for the reason that it is preëminently a dairy breed, surpassing all other pure breeds in the production of rich milk and butter on soils of medium fertility, and admirably adapted, in my opinion, to raise the character of our stock to a higher standard of excellence. The best milkers I have ever known, in the course of my own observations, were grade Ayrshires, larger in size than the pure bloods, but still sufficiently high grades to give certain signs of their origin. I have owned several such, which were all good cows. This grade would seem to possess the advantage of combining, to some extent, the two qualities of milking and adaptation to beef; and this is no small recommendation of the stock to farmers situated as American farmers are, who wish for milk for some years and then to turn over to the butcher.

3

THE JERSEY cattle have now become widely known in this country. Many of them have been imported from an island of the same name in the British Channel, near the coast of France, and they may now be considered, I think, as fully acclimated. They were first introduced over thirty years ago, from the channel islands Alderney, Guernsey, and Jersey.

Fig. 3.   Jersey Cow.*

The opinions of practical men differ widely as to the comparative merits of this race, and its adaptation to our climate and to the wants of our farmers. The most common decision, prevailing among many even of the best judges of stock, appears to be, that, however desirable the cows may be on the lawn or in a gentleman's park, they are wholly unsuited to the general wants of the practical farmer. This may or may not be the case. If the farmer keeps a dairy farm and sells only milk, the quantity and not the quality of which is his chief care, he can satisfy himself better with some other breed. If otherwise situated,—if he devotes his time

* See page 30

to the making of butter for the supply of customers who are willing to pay for a good article,—he may very properly consider whether a few Jerseys, or an infusion of Jersey blood, may not be desirable. Hax ton calls the Jersey cow the cheese and butter dairy-man's friend when her milk is diluted with that of ten or a dozen ordinary cows, and his enemy if he should attempt to make either cheese or butter solely from her produce, as, from the excessive richness of the milk, neither will keep long; and, finally, an ornament to the rich man's lawn, yet in aspect altogether devoid of those rounded outlines which constitute the crite-rion of animal beauty.

The Jersey race is supposed to have been derived originally from Normandy, in the northern part of France. The cows have been long celebrated for the production of very rich milk and cream, but till within a quarter of a century they were comparatively coarse, ugly, and ill-shaped. Improvements have been very marked, but the form of the animal is still far from satisfying the eye. The head of the pure Jersey is fine and tapering, the cheek small, the throat clean. the muzzle fine and encircled with a light stripe, the nostril high and open; the horns smooth, crumpled, not very thick at the base, tapering, and tipped with black; ears small and thin, deep orange color inside; eyes full and placid; neck straight and fine; chest broad and deep; barrel hooped, broad and deep, well ribbed up; back straight from the withers to the hip, and from the top of the hip to the setting on of the tail; tail fine, at right angles with the back, and hanging down to the hocks; skin thin, light color and mellow, covered with fine soft hair; fore legs short, straight and fine below the knee, arm swelling and full above; hind quarters long and well filled; hind legs short and straight below the hocks, with bones rather fine, squarely placed, and not too close together; hoofs small; udder full in size, in line with the belly, extending well up behind; teats of medium size, squarely placed and wide apart, and milk-veins very prominent. The color is generally cream, dun, or yellow, with more or less white, and the fine head and neck give the cows and heifers a fawn-like appearance, and make them objects of attraction in the park; but the hind quarters are often too narrow to look well, particularly to those who judge animals from the amount of fat they carry. We should bear in mind, however, that a good race of animals is not always the most beautiful, as that term is commonly understood. Beauty in stock has no fixed standard. In the estimation of some, it results mainly from fine forms, small bones, and close, compact frames; while others consider that structure the most perfect, and therefore the most beautiful, which is best adapted to the use to which it is destined. According to the latter, beauty is relative. It is not the same in an animal designed for beef and in one designed for the dairy or for work. The beauty of a milch cow is the result of her good qualities. Large milkers are very rarely cows that please the eye of any but a skilful judge. They are generally poor, because their food goes mainly to the production of milk, and because they are selected with less regard to form than to good milking qualities. We meet with good milkers of all forms, from the round, close-built Devon to the coarsest-boned scrub; but, with all their varieties of form and structure, good cows will usually possess certain points of similarity and well-known marks by which they are known to the eyes of the judge.

It is asserted by Colonel Le Couteur, of the island of Jersey, that, contrary to the general opinion here, the

Jersey cow, when old and no longer wanted as a milker, will, when dry and fed, fatten rapidly, and produce a good quantity and excellent quality of butcher's meat. An old cow, he says, was put up to fatten in October, 1850, weighing 1125 pounds, and when killed, the 6th of January, 1851, she weighed 1330 pounds; having gained 205 pounds in ninety-eight days, on twenty pounds of hay, a little wheat-straw, and thirty pounds of roots, consisting of carrots, Swedes, and mangold wurzel, a day. The prevailing opinion as to the beauty of the Jersey is based on the general appearance of the cow in milk, no experiments in feeding exclusively for beef having been made, to my knowledge, and no opportunity to form a correct judgment from actual observation having been furnished; and it must be con-fessed that the general appearance would amply justify the hasty conclusion.

Fig. 4.   Jersey Bull.

The bulls are usually very different in character and disposition from the cows, and are much inclined to

3*

become restive and cross at the age of two or three years, unless their treatment is uniformly gentle and firm. The accompanying figure very accurately represents one of the best animals of the race in the vicinity of Boston, which has been pronounced by good judges a model of a bull for a dairy breed.

The beautiful Jersey cow " Flirt," figured on page 26, received the first prize at the Fair of the Massachusetts State Board of Agriculture in 1857, which brought together the largest and finest collection of Jersey cattle ever made in this country. She is well-shaped, and a very superior dairy cow. Her dam, Flora, was very remarkable for the richness of her milk and the quantity of her butter, having made no less than five hundred and eleven pounds in one year, without extra feeding.

From what has been said it is evident that the Jersey is to be regarded as a dairy breed, and that almost exclusively. It is evident, too, that it would not be sought for large dairies kept for the supply of milk to cities; for, though the quality would gratify the customer, the quantity would not satisfy the owner. The place of the Jersey cow is rather in private establishments, where the supply of cream and butter is a sufficient object, or, in limited numbers, to add richness to the milk of large butter dairies. Even one or two good Jersey cows with a herd of fifteen or twenty, will make a great difference in the quality of the milk and butter of the whole establishment; and they would probably be profitable for this, if for no other object.

Other breeds are somewhat noted in Great Britain for their excellent dairy qualities, and among them might be named the Yorkshire and the Kerry; but they have never been introduced into this country to any extent; though a few of the latter were imported into Massachusetts in 1859, and give very satisfactory results.

Fig. 5 Improved Short-horn DUCHESS. (64.)

THE SHORT-HORNS.—No breed of horned cattle has commanded more universal admiration during the last half-century than the improved Short-horns, whose origin can be traced back for nearly a hundred years. According to the best authorities, the stock which formed the basis of improvement existed equally in Yorkshire, Lincolnshire, Northumberland, and counties adjoining; and the preëminence was accorded to Durham, which gave its name to the race, from the more correct principles of breeding which seem to have prevailed there.

There is a dispute among the most eminent breeders as to how far it owes its origin to early importations from Holland, whence many superior animals were brought for the purpose of improving the old longhorned breed. A large race of cattle had existed for many years on the western shores of the continent of Europe. At a very early date, as early as 1633, they were imported from Denmark into New England in considerable numbers, and thus laid the foundation of a valuable stock in this country. They extended along the coast, it is said, through Holland to France. The dairy formed a prominent branch of farming at a very early date in Holland, and experience led to the greatest care in the choice and breeding of dairy stock. From these cattle many selections were made to cross over to the counties of York and Durham. The prevailing color of the large Dutch cattle was black and white, beautifully contrasted.

The cattle produced by these crosses a century ago were known under the name of "Dutch." The cows selected for crossing with the early imported Dutch bulls were generally long-horned, large-boned, coarse animals, a fair type of which was found in the old "Holderness" breed of Yorkshire,—slow feeders, strong in the shoulder, defective in the fore quarter, and not very profitable for the butcher, their meat being "coarse to the palate and uninviting to the eye." Their milking qualities were good, surpassing, probably, those of the improved short-horns. Whatever may be the truth with regard to these crosses, and however far they proved effective in creating or laying the foundation of the modern improved short-horns, the results of the efforts made in Yorkshire and some of the adjoining counties were never so satisfactory to the best judges as those of the breeders along the Tees, who selected animals with greater reference to fineness of bone and symmetry of form, and the animals they bred soon took the lead, and excited great emulation in improvement.

The famous bull "Hubback," bred by Mr. Turner, of Hurworth, and subsequently owned by Mr. Colling, laid the foundation of the celebrity of the short-horns, and it is the pride of short-horn breeders to trace back to him. He was calved in 1777, and his descendants, Foljambe, Bolingbroke, Favorite, and Comet, permanently fixed the characteristics of the breed. Comet was so highly esteemed among breeders, that he sold at one thousand guineas, or over five thousand dollars. Hubback is thought by some to have been a pure shorthorn, and by others a grade or mixture.

Many breeders had labored long previous to the brothers Charles and Robert Colling, especially on the old Teeswater short-horns; yet a large share of the credit of improving and establishing the reputation of the improved short-horns is generally accorded to the Collings. Certain it is that the spirit and discrimination with which they selected and bred soon became known, and a general interest was awakened in the breed at the time of the sale of Charles Colling's herd, October 11, 1810. It was then that Mr. Bates, of Kirkleavington, purchased the celebrated heifer Duchess I., whose family sold, in 1850, after his decease, at an average of one hundred and sixteen pounds five shillings per head, including young calves. Many representatives of the Duchess family, which laid the foundation of Mr. Bates' success as a breeder, have been brought to this country. They may, perhaps, be regarded as an exception to the modern improved short-horns, their milking qualities being generally very superior.

The sale referred to, and those of R. Colling's herd, in 1818, and that of Lord Spencer, in 1846, as well as that of the Kirkleavington herd, in 1850, and especially that of the herd of Lord Ducie, two years later, are marked eras in the history of improved short-horns; and through these sales, and the universal enthusiasm awakened by them, the short-horns have become more widely spread over Great Britain, and more generally fashionable, than any other breed. They have also been largely introduced into France by the government, for the improvement of the various French breeds by crossing, and into nearly every quarter of the civilized world.

Fig. 6.   Short-horn Bull "DOUBLE DUKE," (1451) Am. H. Book,)
Owned by Harvest Club, Springfield.

Importations have been frequent and extensive into the United States within the last few years, and this famous breed is now pretty generally diffused over the country.

The use of the early-imported short-horn bulls and native cows led to the formation of many families of grades, some of them bred back to the sire, and others crossed high up, which have attained a very consider- able local reputation in many sections. As instances of this, may be mentioned the Creampot stock, obtained by Col. Jaques from a short-horn bull, Coelebs, and a superior native cow. A family of fine milkers still exists in Massachusetts, known by the name of the "Sukey breed," supposed to have been derived from "Denton," a very superior animal imported by Mr. Wil- liams, of Northboro', some forty years age. Many of the best milkers of that section can be traced back to

him. The Patton stock, originally imported into Mary-
land and Virginia, in 1783, and thence to Kentucky,
may be classed in the same category. A part of these
were at first known as the "milk breed," and others as
the "beef breed:" the first short-horns, at that time
good milkers, and the latter long-horns, of large size
and coarse in the bone. In Kentucky they were all
known as the Patton stock.

The high-bred short-horn is easily prepared for a
show, and, as fat will cover faults, the temptation is
often too great to be resisted ; and hence it is common
to see the finest animals rendered unfit for breeding
purposes by over-feeding. The race is susceptible of
breeding for the production of milk, as several families
show, and great milkers have often been known among
pure-bred animals ; but it is more common to find it
bred mainly for the butcher, and kept accordingly. It
is, however, a well-known fact that the dairies of London are stocked chiefly with short-horns and York-
shires, or high grades between them, which, after being
milked as long as profitable, feed equal, or nearly so,
to pure-bred short-horns.

It has been said, by very high authority, that "the
short-horns improve every breed they cross with."

The desirable characteristics of the short-horn bull
may be summed up, according to the judgment of the
best breeders, as follows: He should have a short but
fine head, very broad across the eyes, tapering to the
nose, with a nostril full and prominent; the nose itself
should be of a rich flesh-color ; eyes bright and mild :
ears somewhat large and thin ; horns slightly curved
and rather flat, well set on a long, broad, muscular
neck ; chest wide, deep, and projecting ; shoulders fine,
oblique, well formed into the chine ; fore legs short
with upper arm large and powerful ; barrel round, deep.

well ribbed home; hips wide and level; back straight
from the withers to the setting on of the tail, but short
from hip to chine; skin soft and velvety to the touch:
moderately thick hair, plentiful, soft, and mossy. The
cow has the same points in the main, but her head is
finer, longer, and more tapering, neck thinner and
lighter, and shoulders more narrow across the chine.

The astonishing precocity of the short-horns, their
remarkable aptitude to fatten, the perfection of their
forms, and the fineness of their bony structure, give
them an advantage over most other races when the
object of breeding is for the shambles. No animal of
any other breed can so rapidly transform the stock of
any section around him as the improved short-horn bull.

But it does not follow that the high-bred short-horns
are unexceptionable even for beef. The very exag-
geration, so to speak, of the qualities which make them
so valuable for the improvement of other and less perfect races, may become a fault when wanted for the
table. The very rapidity with which they increase in
size is thought by some to prevent their meat from
ripening up sufficiently before being hurried off to the
butcher. The disproportion of the fatty to the muscular flesh, found in this to a greater extent than in
races coming slower to maturity, makes the meat of the
thorough-bred short-horn, in the estimation of some,
both less agreeable to the taste and less profitable to
the consumer, since the nitrogenous compounds, true
sources of nutriment, are found in less quantity than in
the meat of animals not so highly bred.

But the improved short-horn is justly unrivalled for
symmetry of form and beauty. I have never seen a
picture or an engraving of an animal which gave an
adequate idea of the beauty of many specimens of this
race, especially of the best bred in Kentucky and Ohio.

where many excellent breeders, favored by a climate and pastures eminently adapted to bring the short-horn to perfection, have not only imported extensively from the best herds in England, but have themselves attained a degree of knowledge and skill equalled only by that of the most celebrated breeders in the native country of this improved race.

In sections where the climate is moist and the food abundant and rich, some families of the short-horns may be valuable for the dairy; but they are most frequently bred exclusively for beef in this country, and in sections where they have attained the highest perfection of form and beauty so little is thought of their milking qualities that they are often not milked at all, the calf being allowed to run with the dam.

Fig. 7.   Imported Dutch Cow.

THE DUTCH is a short-horned race of cattle, which, in the opinion of many, as I have already remarked, contributed largely, about a century ago, to build up the Durham or Teeswater stock.   It has been bred with

4

special reference to dairy qualities, and is eminently
adapted to supply the wants of the dairy farmer.

The cow, Fig. 7, was bred in North Holland, and imported by Winthrop W. Chenery, Esq., of Watertown,

Fig. 8.   Imported Dutch Bull.

in 1857.   The bull, Fig. 8, was also imported by Mr.
Chenery at the same time, from near the Beemster, in the
northerly part of Purmerend.   Both animals are truth-
fully delineated, and give a correct idea of the points
of the North Dutch cattle.   For a more detailed description of this celebrated dairy race, see pages 51 and 301.

HEREFORDS. — The Hereford cattle derive their name
from a county in the western part of England.   Their
general characteristics are a white face, sometimes mottled; white throat, the white generally extending back
on the neck, and sometimes, though rarely, still further
along on the back.   The color of the rest of the body
is red, generally dark, but sometimes light.   Eighty
years ago the best Hereford cattle were mottled or
roan all over; and some of the best herds, down to a

comparatively recent period, were either all mottled, or had the mottled or speckled face.   The expression

Fig. 9.   Hereford Cow.

of the face is mild and lively; the forehead open, broad, and large ; the eyes bright and full of vivacity ; the horns glossy, slender, and spreading; the head small though larger and not quite so clean as that of the Devons; the lower jaw fine; neck long and slender; chest deep; breast-bone large, prominent, and very muscular; the shoulder-blade light; shoulder full and soft; brisket and loins large; hips well developed, and on a level with the chine; hind quarters long and well filled in; buttocks on a level with the back, neither falling off nor raised above the hind quarters; tail slender, well set on; hair fine and soft; body round and full; carcass deep and well formed, or cylindrical; bone small; thigh short and well made; legs short and straight, and slender below the knee ; as handlers very excellent, especially mellow to the touch on the back, the shoulder, and along the sides, the skin being soft, flexible, of medium thickness, rolling on the neck and the

hips ; hair bright; face almost bare, which is characteristic of pure-bred Herefords. They belong to the
middle-horned division of the cattle of Great Britain,
to which they are indigenous. They have been improved within the last century by careful selections;
the first step to this end having been taken by Benja-
min Tomkins, of Herefordshire, who began about 1766,
with two cows possessing a remarkable tendency to
take on fat. One of these was gray, and the other
dark red, with a mottled or spotted face.

Taking these as a foundation, Mr. Tomkins went on
to build up a large herd, from which he sold to
other breeders, from time to time, till at his decease, in
1819, the whole herd was disposed of at auction — fifty-
two animals, including twenty-two steers and two heifers, varying in age from calves to two-year-olds, bringing an aggregate of four thousand six hundred and
seventy-three pounds, fourteen shillings, or four hundred and forty-five dollars, thirty-seven and a half cents,
a head. A bull was sold to Lord Talbot for five hundred and eighty-eight pounds, while several cows
brought from a thousand to twelve hundred dollars a
head.

Hereford oxen are excellent animals, less active but
stronger than the Devons, and very free and docile.
The demand for Herefords for beef prevents their being
much used for work in their native county, and the
farmers there generally use horses instead of oxen. A
recent writer in the *Farmer's Magazine* makes the following remarks on this head: "It is allowed on all
hands, I believe, that the properties in which Herefords
stand preëminent among the middle-sized breeds are in
the production of oxen and their superiority of flesh.
On these points there is little chance of their being
excelled. It should, however, be borne in mind that

the best oxen are not produced from the largest cows ; nor is a superior quality of flesh, such as is considered

Fig. 10. Hereford Bull.

very soft to the touch, with thin skin. It is the union of these two qualities which often characterizes the short-horns ; but the Hereford breeders should endeavor to maintain a higher standard of excellence, — that for which the best of the breed have always been esteemed, — a moderately thick, mellow hide, with a well-appor- tioned combination of softness with elasticity. A suffi- ciency of hair is also desirable, and if accompanied with a disposition to curl moderately it is more in esteem; but that which has a harsh and wiry feel is objection- able."

In point of symmetry and beauty of form, the well- bred Herefords may be classed with the improved short- horns, though they arrive somewhat slower at maturity, and never attain such weight. Like the improved short- horns, they are chiefly bred for beef, and their beef is of the best quality in the English markets, command- ing the highest price of any, except, perhaps, the West Highlanders.

4*

In an experiment carefully tried in 1828, for the purpose of arriving at the comparative economy of the
short-horns and Herefords, the latter gained less by
nearly one fourth than the former, which had consumed
more food.   The six animals, three of each breed, were
sold after being fed, in Smithfield market, the Herefords
bringing less by only about five dollars than the short-
horns, while the cost of food consumed by the latter
was greater, and the original purchase greater than that
of the former.

The short-horn produces more beef at the same age
than the Hereford, but consumes more food in proportion.   "In all the fairs of England," says Hillyard, " except those of Herefordshire and the adjoining counties,
short-horn heifers are more sought after and sell at
higher prices than the Hereford; but it is not so with
fat cattle, for, with the exception of Lincolnshire and
some of the northern counties, they much prefer the
Herefords.   Then at Smithfield, where the quality of
the beef passes its final judgment, the pound of Hereford beef pays better than the pound of short-horn beef.
Short-horn beeves produce at the same age a greater
weight, it is true, but they also consume more food.   I
can easily conceive why, in the magnificent pastures of
Lincolnshire, and some of the northern counties of
England, they may prefer the short-horns; and that is,
that they may keep a less number on a given quantity
of land, and only the short-horn could, under these conditions, produce a greater weight of beef per acre.   It
is very difficult to decide which of the two races in
England (the two best in the world) is the most profitable for stock-raisers and for the community."   There
are, even in Lincolnshire, many good feeders who prefer the Herefords to the short-horns.   One of these,
when visited the past season, had thirty head of cattle

feeding for the butcher, and only one short-horn. When asked the reason of this, he replied, " I am a farmer myself, and have to pay high rent, and I must feed the cattle that pay me best. Perhaps you think it would be more in fashion to cover my fields with short-horns; but I must look to the net profit, and I get much better with the Herefords. The short-horns are too full of fat and make too little tallow, and they consequently sell too low in the Smithfield market. Our Herefords are better, and they sell better."

The Herefords are far less generally spread over England than the improved short-horns. They have seldom been bred for milk, as some families of the short-horns have; and it is not very unusual to find pure-bred cows incapable of supplying milk sufficient to nourish their calves. This system was pursued especially by Mr. Price, a skilful Hereford breeder, who sacrified everything to form, disregarding milking properties, breeding often from near relations, and thus fixing the fault incident to his system more or less permanently in the descendants of his stock.

The Herefords have been brought to this country, to some extent, and several fine herds exist in different sections; the earliest importations being those of Henry Clay, of Kentucky, in 1817. The figures of the two animals of this breed presented in this connection represent a bull and cow bred at the State Farm, in Massachusetts, and are good specimens of the breed.

The want of care and attention to the udder, soon after calving, especially if the cow be on luxuriant grass, often injures her milking properties exceedingly. The practice in the county of Hereford has generally been to let the calves suckle from four to six months, and bull-calves often run eight months with the cow. But their dairy qualities are perhaps as good as those of any cattle whose fattening properties have been so carefully developed; and, though it is probable that they could be bred for milk by proper care and attention, yet, as this change would be at the sacrifice of other qualities equally valuable, it would evidently be wiser to resort to other stock for the dairy.

Fig. 11   Devon Cow.
Owned by William Buckminster Esq., Framingham Mass.

THE NORTH DEVONS. — The last of the pure-bred races which it will be necessary to describe as prominent among our American cattle is the Devon, a middle-horned breed, now very generally distributed in some sections of the country.

This beautiful race of cattle dates further back than any well-established breed among us. It goes generally under the simple name of Devon; but the cattle of the southern part of the county, from which the race derives its name, differ somewhat from those of the northern, having a larger and coarser frame, and far less tendency to fatten, though their dairy qualities are superior.

The North Devons are remarkable for hardihood, symmetry, and beauty, and are generally bred for work and for beef rather than for the dairy. The head is fine and well set on; the horns of medium length, generally curved; color usually bright blood-red, but sometimes inclining to yellow; skin thin and orange-yellow; hair of medium length, soft and silky, making the animals remarkable as handlers; muzzle of the nose white; eyes full and mild; ears yellowish, or orange-color inside, of moderate size; neck rather long, with little dewlap; shoulders oblique; legs small and straight, and feet in proportion; chest of good width; ribs round and expanded; loins of first-rate quality, long, wide, and fleshy; hips round, of medium width; rump level; tail full near the setting on, tapering to the tip; thighs of the bull and ox muscular and full, and high in the flank, though in the cow sometimes thought to be too light; the size medium, generally called small. The proportion of meat on the valuable parts is greater, and the offal less, than on most other breeds, while it is well settled that they consume less food in its production. The Devons are popular with the Smithfield butchers, and their beef is well marbled or grained.

As working oxen, the Devons perhaps excel all other races in quickness, docility, and beauty, and the ease with which they are matched. With a reasonable load, they are said to be equal to horses as walkers on the road, and when they are no longer wanted for work they fatten easily and turn well.

As milkers, they do not excel, perhaps they may be said not to equal, the other breeds, and they have a reputation of being decidedly below the average. In their native country the general average of a dairy is one pound of butter per day during the summer.

They are bred for beef and for work, and not for the

dairy; and their yield of milk is small, though of a rich quality. I have, however, had occasion to examine several animals from the celebrated Patterson herd, which would have been remarkable as milkers even among good milking stock. They had not, to be sure, the beautiful symmetry of form and fineness of bone which characterize most of the modern and highly improved pure-bred North Devons, and had evidently been bred for many years with special reference to the development of the milking qualities, great care having been taken to use bulls and cows as breeders from the best milking stock, rather than of the finest forms. The use of bulls distinguished only for symmetry of form, and of a race deficient in milk-secreting qualities, will be sure to deteriorate, instead of improving, the stock for the dairy.

Fig. 12. Devon Bull.

On the whole, whatever may be our judgment of this breed, the faults of the North Devon cow can hardly be overlooked from our present point of view. The rotundity of form and compactness of frame, though they contribute to her remarkable beauty, constitute an objection to her as a dairy cow, since it is generally thought that the peculiarity of form which disposes an animal to take on fat is somewhat incompatible with good milking qualities, and hence Youatt says: " For the dairy the North Devons must be acknowledged to be inferior to several other breeds. The milk is good, and yields more than the average proportion of cream and butter; but it is deficient in quantity." He also maintains that its property as a milker could not be improved without probable or certain detriment to its grazing qualities.

But the fairest test of its fitness for the dairy is to be found in the estimation in which distinguished Devon breeders themselves have held it in this respect. A scale of points of excellence in this breed was established, some time ago, by the best judges in England; and it has since been adopted, with but slight changes, in this country.. These judges, naturally prejudiced in favor of the breed, if prejudiced at all, made this scale to embrace one hundred points, no animal to be regarded as perfect unless it excelled in all of them. Each part of the body was assigned its real value in the scale: a faultless head, for instance, was estimated at four; a deep, round chest, at fifteen, &c. If the animal was defective in any part, the number of points which represented the value of that part in the scale was to be deducted pro rata from the hundred, in determining its merits. But in this scale the cow is so lightly esteemed for the dairy, that the udder, the size and shape of which is of the utmost consequence in determining the capacity of the milch cow, is set down as worth only *one point*, while, in the same scale, the horns and ears are valued at two points each, and the color of the nose, and the expression of the eye, are valued at four points each. Supposing, therefore, that each of these points were valued at one dollar, and a perfect North Devon cow was valued at one hundred dollars ; then another cow of the same blood, and equal to the first in every respect except in her udder, which is such as to make it certain that she can never be capable of giving milk enough to nourish her calf, must be worth, according to the estimation of the best Devon breeders, ninety-nine dollars ! It is safe, therefore, to say that an animal whose udder and lacteal glands are regarded, by those who best know her capacities and her merits, as of only one quarter part as much consequence as the color of her nose, or half as much as the shape and size of her horns, cannot be recommended for the dairy. The improved North Devon cow may be classed, in this respect, with the Hereford, neither of which has well-developed milk-vessels — a point of the utmost consequence to the practical dairyman.

The list of pure-bred races in America may be said to end here ; for, though other and well-established breeds, like the long-horns, the Galloways, the Spanish, &c., have, at times, been imported, and have had some influence on our American stock, they have not been kept distinct to such an extent as to have become the prevailing stock of any particular section, so far as I am aware, and hence a notice of them properly comes in the next chapter.

CHAPTER II.

AMERICAN GRADE OR NATIVE CATTLE.—THE PRIN-
CIPLES OF BREEDING.

WE have dwelt thus far mainly upon the prominent breeds of cattle known among us, and especially those adapted to the dairy. But a large proportion—by far the largest proportion, indeed—cannot be included under any of the races alluded to.

The term breed, properly understood, applies only to animals of the same species, possessing, besides the general characteristics of that species, other characteristics peculiar to themselves, which they owe to the influence of soil, climate, nourishment, and habits of life to which they are subjected, and which they transmit with certainty to their progeny. The characteristics of certain breeds or families are so well marked, that if an individual supposed to belong to any one of them were to produce an offspring not possessing them, or possessing them only in part, with others not belonging to the breed, it would be just ground for suspecting a want of purity of blood.

If this definition of the term breed be correct, no grade animals, and no animals not possessing fixed peculiarities or characteristics which they share with all other animals of the class of which they are a type, and which they are capable of transmitting with certainty to their descendants, can be recognized by breeders as belonging to any one distinct race, breed, or family.

The term " native," or " scrub," is applied to a vast
majority of our American cattle, which, though born on
the soil, and thus in one sense natives, do not constitute
a breed, race, or family, as properly understood by
breeders. They do not possess characteristics peculiar
to them all, which they transmit with any certainty to
their offspring, either of form, size, color, milking or
working properties. But, though an animal may be
made up of a mixture of blood almost to infinity, it does
not follow that, for specific purposes, it may not, as an
individual animal, be one of the best of the species.
And for particular purposes individual animals might
be selected from among those commonly called natives
in New England, and scrubs at the West and South,
equal, and perhaps superior, to any among the races
produced by the most skilful breeding. There can be
no impropriety in the use of the term " native," therefore, when it is understood as descriptive of no known
breed, but only as applied to the common stock of the
country, which does not constitute a breed. But perhaps the whole class of animals commonly called " natives " would be better described as grades, since they
are well known to have sprung from a great variety of
cattle procured in different places and at different times
on the continent of Europe, in England, and in the
Spanish West Indies, brought together without any
regard to fixed principles of breeding, but only from
individual convenience, and by accident.

The first importations to this country were doubtless
those taken to Virginia previous to 1609, though the
exact date of their arrival is not known. Several cows
were carried there from the West Indies in 1610, and
the next year no less than one hundred arrived there
from abroad.

The earliest cattle imported into the Plymouth colony, and undoubtedly the earliest introduced into New England, arrived in 1624. At the division of cattle which took place in 1627, three years after, one or two are distinctly described as black, or black and white, others as brindle, showing that there was no uniformity of color. Soon after this, a large number of cattle were brought over from England for the settlers at Salem. These importations formed the original stock of Massachusetts.

In 1625 the first importation was made into New York from Holland, by the Dutch West India Company, and the foundation was then laid for an exceedingly valuable race of animals, which subsequent importations from the same country, as well as from England, have greatly improved.

Dairy farming in some parts of Holland, it may be remarked in passing, became a highly important branch of industry at a very early date, and a large and valuable race of dairy cattle existed there long before the efforts of modern breeders began in England. The attention of farmers there is at the present time devoted especially to the dairy, and the manufacture of butter and cheese. They support themselves, to a considerable extent, upon this branch of farming; and hence it is held in the highest respect, and carried to a greater degree of exactness and perfection, perhaps, than in any other part of the world. They are especially particular in the breeding, keeping, and care of milch cows, as on them very much of their success depends. The principles on which they practise, in selecting a cow to breed from, are as follows: She should have, they say, considerable size — not less than four and a half or five feet girth, with a length of body corresponding; legs proportionally short; a finely-formed head, with a forehead or face somewhat concave; clear,

large, mild, and sparkling eyes, yet with no expression
of wildness; tolerably large and stout ears, standing out
from the head; fine, well-curved horns; a rather short
than long, thick, broad neck, well set against the chest
and withers; the front part of the breast and the shoulders must be broad and fleshy; the low-hanging dewlap
must be soft to the touch; the back and loins must be
properly projected, somewhat broad, the bones not too
sharp, but well covered with flesh; the animal should
have long, curved ribs, which form a broad breast-bone;
the body must be round and deep, but not sunken into
a hanging belly; the rump must not be uneven, the hip-
bones should not stand out too broad and spreading,
but all the parts should be level and well filled up; a
fine tail, set moderately high up and tolerably long, but
slender, with a thick, bushy tuft of hair at the end,
hanging down below the hocks; the legs must be short,
and low, but strong in the bony structure; the knees
broad, with flexible joints; the muscles and sinews must
be firm and sound, the hoofs broad and flat, and the
position of the legs natural, not too close and crowded;
the hide, covered with fine glossy hair, must be soft and
mellow to the touch, and set loose upon the body. A
large, rather long, white and loose udder, extending
well back, with four long teats, serves also as a characteristic mark of a good milch cow. Large and prominent milk-veins must extend from the navel back to
the udder; the belly of a good milch cow should not be
too deep and hanging. The color of the North Dutch
cattle is mostly variegated. Cows with only one color
are no favorites. Red or black variegated, gray and
blue variegated, roan, spotted and white variegated
cows, are especially liked.

The annexed cut represents a cow most esteemed in the
North of France. It is the type of the race so noted for

the production of milk, and of the excellent dairy breeds of Holland and the low countries.

In 1627, cattle were brought from Sweden to the settlements on the Delaware by the Swedish West India Company. In 1631, 1632, and 1633, several importa-

Fig. 13. Dutch Dairy Cow.

tions were made into New Hampshire by Capt. John Mason, who, with Gorges, procured the patent of large tracts of land in the vicinity of Piscataqua River, and immediately formed settlements there. The object of Mason was to carry on the manufacture of potash. For this purpose he employed the Danes; and it was in his voyages to and from Denmark that he procured many Danish cattle and horses, which were subsequently diffused over that whole region, and large numbers of which were driven to the vicinity of Boston and sold. These facts are authenticated by original documents and depositions now on file in the office of the Secretary of State of New Hampshire. The Danish cattle are there described as large and coarse, of a yellow color; and it is supposed that they were procured by

5*

Mason as being best capable of enduring the severity of the climate and the hardships to which they were to be subjected. However this may have been, they very soon spread among the colonists of the Massachusetts Bay, and have undoubtedly left their marks on the stock of New England and the Middle States, which exist to some extent even to the present day, mixed in with an infinite multitude of crosses with the Devons, the Dutch cattle already alluded to, the black cattle of Spain and Wales, and the long-horn and the short-horn, most of which crosses were accidental, or due to local circumstances or individual convenience. Many of these cattle, the descendants of such crosses, are of a very high order of merit, but to what particular cross it is due it is impossible to say. They make generally hardy, strong, and docile oxen, easily broken to the yoke and quick to work, with a fair tendency to fatten when well fed; while the cows, though often ill-shaped, are sometimes remarkably good milkers, especially as regards the quantity they give.

I have very often heard the best judges of stock say that if they desired to select a dairy of cows for milk for sale, they would go around and select cows commonly called native, rather than resort to pure-bred animals of any of the established breeds, and that they believed they should find such a dairy the most profitable.

In color, the natives, made up as already indicated, are exceedingly various. The old Denmarks, which to a considerable extent laid the foundation of the stock of Maine and New Hampshire, were light yellow. The Dutch of New York and the Middle States were black and white; the Spanish and Welsh were generally black; the Devons, which are supposed to have laid the foundation of the stock of some of the states, were red. Crosses of the Denmark with the Spanish and Welsh naturally made a dark brindle. Crosses of the Denmark and Devon often made a lighter or yellowish brindle, while the more recent importations of Jerseys and short-horns have generally produced a beautiful spotted progeny. The deep red has long been a favorite color in New England; but the prejudice in its favor is fast giving way to more variegated colors.

But, though we have already an exceedingly valua- ble foundation for improvement, no one will pretend to deny that our cattle, as a whole, are susceptible of it in many respects. They possess neither the size, the sym- metry, nor the early maturity, of the short-horns; they do not, as a general thing, possess the fineness of bone, the beauty of form and color, nor the activity, of the Devons or the Herefords; they do not possess that uniform richness of milk, united with generous quantity, of the Ayrshires, nor the surpassing richness of milk of the Jerseys; but, above all, they do not pos- sess the power of transmitting the many good qualities which they often have to their offspring, which is a characteristic of all well-established breeds.

Equally certain is it, in the opinion of many good judges, that the dairy stock of New England has not been improved in its intrinsic good qualities during the last thirty or forty years. Cows of the very highest order as milkers were as frequently met with, they say, in 1825, as at the present time. Any increased product of our dairies they ascribe to improved care and feed ing, rather than to improvement in the dairy qualities of the stock.

This may not be true of some other sections of the country, where the dairy has been a more special object of pursuit, and where the custom of raising the best male calves of the neighborhood, or those that came from the best dairy cows, and then of using only the best-formed bulls, has long prevailed. In this way some progress has, doubtless, been made.

There is an old adage among the dairy farmers of Ayrshire, that "The cow gives her milk by the mou'," which was slightly varied from an old German proverb, that "*The cow milks only through the throat.*" It is fortunate, indeed, that wiser and more humane ideas prevail with regard to the care of stock of all kinds; for it is well known that the treatment the stock of the country received for the first two centuries after its settlement was often barbarous and cruel in the extreme, and that thousands perished, in the early history of the colonies, from exposure and starvation. Even within my own distinct recollection, it was thought, for miles around my native place, that cows and young stock should remain out of doors exposed to the cold winter days, to "toughen;" and that, too, by men who styled themselves "practical" farmers.

Mr. Henry Colman truly asserted, in 1841, that the general treatment of cows in New England would not be an inapt subject of presentment by a grand jury. There were, at that time, it is true, many honorable exceptions; but the assertion was strictly correct so far as it applied to the section of which I then had a personal knowledge. Judging from the anxiety manifested by those who enter superior milch cows for the premiums offered by agricultural societies to show that they have had nothing, or next to nothing, to eat, it is evident that the false ideas with regard to the feeding and treatment of this animal have not yet wholly disappeared. But, if little improvement has been made in our dairy stock except that produced by more liberal feeding, it simply shows that our efforts have not been made in the right direction.

The raising of cattle has now become a source of

profit in many sections to a greater extent, at least, than formerly, and it becomes a matter of great practical importance to our farmers to take the proper steps to improve them. Indeed, the questions, what is the best breed, and what are the best crosses, and how shall I improve my stock, are now almost daily asked; and their practical solution would add many thousand dollars to the aggregate wealth of the farmers of the country, if they would all study their own interests. The time is gradually passing away when the intelligent practical farmer will be willing to put his cows to any mere "runt" of a bull, simply because his service may be had for twenty-five cents; for, even if the progeny is to go to the butcher, the calf sired by a pure-bred bull, particularly of a race distinguished for fineness of bone, symmetry of form, and early maturity, will bring a much higher price at the same age than the calf sired by a scrub. Blood has a money value, which will, sooner or later, be generally appreciated. The first and most important object of the farmer is to get the greatest money-return for his labor and his produce; and it is for his interest to obtain an animal — a calf, for instance — that will yield the largest profit on the outlay. If a calf, for which the original outlay was five dollars, will bring at the same age, and on the same keep, more real net profit than another, the original outlay for which was but twenty-five cents, it is certainly for the farmer's interest to pay the larger original outlay, and have the superior animal. Setting all fancy aside, it is merely a question of dollars and cents; but one thing is certain, and that is, that the farmer cannot afford to keep poor stock. It eats as much, and requires nearly the same amount of care and attention, as stock of the best quality; while it is equally certain that stock of ever so good a quality, whether grade, "native," or thorough-bred, will be sure to deteriorate and sink to the level of poor stock, by neglect and want of proper attention.

How, then, are we to improve our stock? Not, surely, by that indiscriminate crossing, with a total disregard to all well-established principles, which has thus far marked our efforts generally with foreign stock, and which is one prominent reason why so little improvement has been made in our dairies; nor by leaving all the results to chance, when, by a careful and judicious selection, they may be within our own control. Two modes of improvement seem to suggest themselves to the mind of the breeder, either of which, apparently, promises good results. The first is, to select from among our native cattle the most perfect animals not known or suspected to be related to any of the well-established breeds, and to use them as breeders. This is a mode of improvement simple enough, if adopted and carried on with animals of any known breed; and, indeed, it is the only mode of improvement which preserves the purity of blood; but, to do it successfully, requires great experience, a good and sure eye for stock, a mind free from prejudice, and indefatigable patience and perseverance. It is absolutely necessary, also, to pay special attention to the calves thus produced; to furnish them at all times, summer and winter, with an abundant supply of nutritious food, and to regulate it according to their growth. Few men are to be found willing to undertake the herculean task of building up a new breed in this way from grade stock. An objection meets us at the very outset, which is that it would require a long series of years to arrive at any satisfactory results, from the fact that no two animals, made up, as our "native" cattle are, of such a variety of elements and crosses, could be found sufficiently alike to produce their kind. The principle that like produces like may be perfectly true, and in the well-known breeds it is not difficult to find two animals that will be sure to transmit their own characteristics to their offspring; but, with two animals which cannot be classed with any breed, the defects of an ill-bred ancestry will be liable to appear through several generations, and thus thwart and disappoint the expectations of the breeder. The objection of time, and expense, and disappointment, attending this method, should have no weight, if there were no* more speedy method of accomplishing equally desirable results.

The second mode is somewhat more feasible; and that is, to select animals from races already improved and well-nigh perfected, to cross with our cattle, using none but good specimens of pure-bred males, and selecting, if our object is to improve stock for the dairy, only such as belong to a race distinguished for dairy qualities; or, if resort is had to other breeds less remarkable for such qualities, such only as are descended from large and generous milkers. And here it may be remarked that these qualities do not belong to any one breed exclusively, though, as they depend mainly on structure and temperament, which are hereditary to a considerable extent, they are themselves transmissible. In almost every breed we can find individual good milkers which greatly surpass the average of the cows of the same race or family, and from such many suppose that good crosses may be expected. How often do we see farmers raising the calves of their best milking-cows simply because they are the best cows, without regard to the qualities of the bull, or to the progenitors of either parent; and how often are they disappointed, at the end of three or four years of labor and expense! Now, though a cow of a bad milking family, or of a breed not at all distinguished for dairy qualities, may turn out to

be an excellent milker, and all else that may be desirable
in a cow, yet these qualities in her are accidental. They
are not supposed to be transmissible with anything like
the certainty which exists where they are the fixed
and constant characteristics of the family. She is an
exception to the rule of her race. A good calf from
her, though not, of course, an impossibility, would be
very much the result of chance. The resort to any
but a distinguished breed of milkers cannot, therefore,
be recommended, nor can we expect to improve our
dairies by it. A disregard of this important matter has
led to endless disappointment, and has done much to
raise up unjust prejudices against the use of all improved stock on our native cows. As if we could
expect nature to go out of her regular course to give
us a good animal, when we have violated her laws!

The offspring of these crosses will be grades; but
grades are often better for the practical purposes of
the farmer than pure-bred animals. The skill of the
breeder is especially manifest in the selection of animals
to breed from, since both parents undoubtedly have a
great influence in transmitting the milking qualities of the
race. But this method of improvement requires less
exact and critical knowledge than the first, from the fact
that it is easier to appreciate the good points of an animal already perfected, or greatly improved, than to discover them in animals which it is our desire to improve,
and which are inferior in form, possessing only the elements of a better stock. It has also an immense advantage, since results may be far more rapidly attained, and
improvements effected which, by the first method,—that
of creating or building up a race from the so-called
natives, by judicious selections, — would be looked for
in vain in the ordinary life of man. All grades are produced by this second method; but all grades are not

equally good, nor equally well adapted to meet the farmer's wants.   It is desirable to know, then, what, on the whole, are the best and most profitable to the practical farmer.

We want cattle for distinct purposes, as for milk, beef, or labor.   In a large majority of cases,—especially in the dairy districts, comprising the Middle and Eastern States, at least,—the farmer cares more for the milking qualities of his cows, especially for the quantity they give, than for their fitness for grazing, or aptness to fatten.   These latter points become more important in the Western and some of the Southern States, where far greater attention is paid to breeding and to feeding, and where comparatively little attention is given to the productions of the dairy.   A stock of cattle that might suit one farmer might be wholly unsuited to another; and in each particular case the breeder should have some special object in view, and select his animals with reference to it.   But there are some general principles that apply to breeding everywhere, and which, in many cases, are not well understood.

It would not be desirable, even if it were possible, by crossing, to breed out all the general characteristics of many of our native cattle.   They have many valuable qualities adapted to our climate and soil, and to the geological structure of the country; and these should be preserved, while we improve the points in which many of them are deficient, such as a want of precocity and aptitude to fatten, where it is an object to attain this quality, coarseness of bone, and lack of symmetry, which is often apparent, especially when the form of the animal does not indicate a near relation to some of the established breeds.

It is a well-known fact that, in crossing, the produce

6

most frequently takes after the male parent, especially, it is thought, in exterior form, in its organs of locomotion, such as the bones, the muscles, &c. Particularly is this the case when the male belongs to an old and well-established breed, and the female belongs to no known breed, and has no strongly-marked and fixed points. Put a Galloway bull, for instance, to a native cow, and the calf will, as a general rule, be hornless. Put a ram without horns to ewes with horns, and most of the lambs will be destitute of horns; that is, they take the characteristics of the father rather than the dam; and this rule holds good generally in breeding, though, like all other rules, it has, of course, its exceptions. Hence, if this position be correct, the first principle which the good sense of the farmer would dictate would be to select a bull from a breed most noted for the qualities he wishes to obtain in their greatest perfection, and especially if the cow is deficient in those qualities. A bull, for instance, of fine bone, and other good points in perfection, will make up for the deficiency of some of these points in the cow.

On the other hand, say the advocates of this doctrine, in the physiology of breeding, the internal structure of the offspring, the organs of secretion, the mucous membranes, the respiratory organs, &c., are imparted chiefly by the dam. Hence it has sometimes been found that by taking a cow remarkable for milking properties, though deficient in many other points, as in the coarseness of bone and in early maturity, and putting to her a bull remarkable for symmetry of form and fineness of bone, the offspring has been superior to the cow in beauty of form and proportions, and has still retained the milking qualities of the dam. This principle, as already intimated, is questioned by some, who say that the milking qualities, as well as the external form, &c., are transmitted through the male offspring.

Mr. James Dickson, an experienced breeder and drover, who views the subject from his own standpoint, says : " A great part of the art of breeding lies in the principle of *judicious crossing;* for it is only by attending properly to this that success is to be attained, and animals produced that shall yield the greatest amount of profit for the food they consume. All eminent breeders know full well that ill-bred animals are unprofitable both to the breeder and feeder. To carry out the system of crossing judiciously, certain breeds of cattle, sheep, pigs, &c., must be kept pure of their kind — males especially ; indeed, as a general rule, no animal possessing spurious blood, or admixture with other breeds, should be used. The produce in almost all cases assimilates to the male parent ; and I should say that in crossing the use of any males not pure-bred is *injudicious,* and ought to be avoided."

If, therefore, a cross is effected with satisfactory results, it should be continued by resorting to pure-bred bulls, and not by the use of any grade bulls thus obtained ; for, though a grade bull may be a very fine animal, it has been found that he does not transmit his good qualities with anything like the certainty of a pure-bred one. The more desirable qualities are united in the bull, the better ; but the special reason for the use of a pure-bred male in crossing is not so much that the particular individual selected has these qualities most perfectly developed in himself, as that they are *hereditary in the breed* to which he belongs. The moment the line is crossed, and the pedigree broken, uncertainty commences. Although the form of the grade bull may, in individual cases, be even superior to that of his pure-bred sire, yet there is less likelihood of his transmitting the qualities for which his breed is most noted; and when it is considered that during his life he may scatter his progeny over a considerable section of country, and thus affect the cattle of his whole neighborhood, attention to this becomes a matter of no small public importance.

This principle, so far as its application to breeding for the shambles is concerned, seems to me to be sound, and fully established by long experience and practice. Perhaps it is equally so, also, in breeding for the dairy. But it may be well to consider whether there are not other rational modes of judgment in the selection of animals for breeding with this specific object in view.

There is a difference of opinion with regard to the practical value of the system of classification and judgment of milch cows discovered and developed by Guénon: some being inclined to ridicule it, as absurd; others to adopt it implicitly, and follow it out in all its details; and others still—and among this class I generally find a very large number of the most sensible practical judges of stock — to admit that in the main it is correct, though they discredit the practicability of carrying it so far, and so minutely into detail, as its author did.

It may be remarked, at the outset, that the fact that the best of the signs of a great and good milker adopted by Guénon are generally found united with the best forms and marks almost universally admitted and practised upon by good judges, gives, at least, some plausibility to the system, while the importance of it, if it be correct, is sufficient to demand a careful examination. Every good judge of a milch cow, for instance, wants to see in her a small, fine head, with short and yellowish horns; a soft, delicate, and close coat of hair

a skin soft and flexible over the rump; broad, well-
spread ribs, covered with a loose skin of medium thickness; a broad chest; a long, slender tail; straight
hind legs; a large, regularly-formed udder, covered with
short, close, silky hair; four teats of equal size and
length, set wide apart; large, projecting lacteal veins,
which run along under the belly from the udder towards the fore legs, forming a fork at the end, and
finally losing themselves in a round cavity; and when
these points, or any considerable number of them, are
found united in a cow, she would be pronounced a
good milker.   An animal in which these signs are
found would rarely fail of having a good " milk-mirror,"
or *escutcheon;* on which Guénon, after many years of
careful observation and experiment, came to lay particular stress; and on the basis of which he built up a
system or theory so complicated as to be of little practical value compared with what it might have been had
he seen fit to simplify it so as to bring it within the
easy comprehension of the farmer.   As one means of
forming a judgment of the milking qualities, however,
it must be regarded as very important, since it is unquestionably sustained by facts in a very large majority
of cases.

The milk-mirror, or escutcheon, is formed by the hair
above the udder, extending upwards between the
thighs, growing in an opposite direction from that of
other parts of the body.   In well-formed mirrors, found
only in cows which have the arteries which supply the
milky glands large and fully developed, it ordinarily begins between the four teats in the middle, and ascends
to the vulva, and sometimes even higher, the hair growing upwards.   The direction of the hair is subordinate
to that of the arteries; for the relation existing between
the direction of the hair above the udder and the

6*   5

activity of the milky glands is apparent on a careful examination of all the cases. When the lower part of the mirror is large and broad, with the hair growing from below upwards, and extending well out on the thighs, it indicates that the arteries which supply the milky glands, and which are situated just behind it, are large and capable of conveying much blood, and of giving great activity to the functions of secretion.

Now, in the bull, the arteries which correspond to the mammary or lacteal arteries of the cow are not so fully developed; and the escutcheons are smaller, shorter, and narrower. Guénon applied the same name, milk-mirror, to these marks in the bull; and the natural inference was, that there should exist a correspondence or similarity in the mirror of the bull and the cow which are coupled for the purpose of producing an offspring fit for the dairy, — that the mirror in the bull should be of the same class, or of a better class than that of the cow.

It is confidently asserted by the advocates of Guénon's method, and with much show of reason, that the very large proportion of cows of bad or indifferent milking qualities, compared with the good, is owing to the mistakes in selecting bulls without reference to the proper marks or points. As to the transmission of the milk-mirror, it has been found in many cases that bulls sprung from cows with good mirrors had smaller and more heart-shaped mirrors, spreading out pretty broad upon the thighs. Pabst, a successful German breeder, says that he has used such bulls for three years, and that the milk-mirrors were transmitted in the majority of the male progeny, and in nearly every case very large and beautiful mirrors were given to the heifer-calves. A son of the bull with which he began was serving at the time of which he speaks, having a mirror more highly developed than his sire, and the

first calves of his get had also very large milk-mirrors. The female offspring of the first bull of good milk-
mirror promised first rate, though they had not then
come in. His inference is, that in breeding from cows
noted as milkers regard should be had to the form
of the mirror on the bull, and the chance of his
transmitting it. If any credit is due to this ingenious method, it may be laid down, as a principle in
the selection of a bull to get dairy stock, that the one
possessing the largest and best-developed milk-mirror
is the best for the purpose, and will be most likely to
get milkers of large quantity and continued flow. This
method will be more fully developed in the chapter on
the Selection of Milch Cows.

But, however careful we may be to select good
milkers, and to breed from them with the hope of improvement, it is by no means easy to select such as are
capable of transmitting their qualities to their offspring. This is rendered still more difficult by the
fact that there is no known mark to indicate it, and we
are left to use our own judgment ; for, in the case of
bulls, we are often obliged to give them up before their
progeny have arrived at an age to show their qualities
by actual trial. We are thrown back, therefore, upon
their external marks. But, as M. Magne, a very sensible
French writer, justly observes in his admirable little
work (*Choix des Vaches Latières*, p. 86, Paris, 1857 ,
the fixed characteristics which have existed in races
for several generations will be transmitted with most
certainty. Hence the importance, he says, of selecting
milch cows from good breeds and good families, and
especially, in breeding stock, of selecting carefully both
male and female. The male designed to get dairy stock
ought to possess the structure which, in the cow, indicates the greatest activity of the mammary glands, as

fineness of form, mellowness of skin, large hind quarters, large and well-developed veins and escutcheon.

A cow of a race or family not noted as milkers may chance to be an excellent milker, and this is enough, if we do not desire to breed from her; but she would not transmit her exceptional qualities like a cow of which these qualities were the fixed characteristics, constant and transmissible in the breed. These considerations apply also, as already said, in the choice of a bull. The attention of practical men has been so much directed to the best points of good cows, of late years, that it becomes necessary to study to propagate these, if the breeder desires to find buyers for his stock. The buyer judges more from external signs than from the intrinsic qualities of the cow, with which he may not be acquainted.

To explain the variations in the transmission of milking qualities, we should bear in mind that these qualities are not found in wild cows, and that they are developed only when man can, by a particular course of treatment, as by the act of milking, the separation of the sexes, etc., cause certain natural powers to act with greater strength than others; that they incline to disappear as soon as these powers, the nature of the soil, the peculiarities of climate, the properties of plants, and the temperament of the cows, are permitted to act according to the original plan of creation; so that the variations which we consider as sports of nature are incontestible proofs of the uniformity of her works.

It is only by observing animals carefully, by noting accurately their good qualities and their faults, by watching the circumstances in which individuals are produced, raised, and kept, that we can account for what seems to us a sport or caprice of nature. We can then tell, first, how the same bull and cow have produced three calves with different properties ; and, secondly, trace out the rules which we are to follow, to be almost uniformly successful in obtaining stock of the best quality.

Experience shows that the qualities which are trans-mitted with most certainty depend on the most import-ant organs of life ; and so, in the forms of the viscera and the skeleton, variations are rare, not only in breeds of the same species, but in different species of the same genera.

Moreover, in cases where the transmission of proper-ties is so uncertain as to seem the result of caprice in nature, these properties are formed by superficial organs, — by the skin, the horns, the state of the hair, etc.

But it is in qualities which are, in a measure, arti-ficial, qualities produced by domestication, and often more injurious than useful to the health of animals, that variations most commonly occur. These change not only with the breed of one species, but with the dif-ferent individuals of the same breed, of the same half-breed, and often of the same family.

Bearing these elementary principles of natural his-tory and physiology in mind, we shall comprehend how cows and bulls well marked in regard to escutcheons have produced stock which did not resemble them. M. Lefebvre Sainte Marie asserts that the influence of the escutcheons is very feeble in the act of reproduction.

In this view, the escutcheon is almost nothing in itself. It depends on the state of the hair, on one of the most fleeting of peculiarities, on that which is least hereditary in animals. It has no value as a mark of good getters of stock, unless it is supported by marks superior to it from their stability, — a larger skeleton, double loins, a wide rump, highly-developed blood-vessels, — unless it is united with a spacious chest, round ribs, large lungs, and a strong constitution. The more complete the correspondence between these marks, the more the milking quality is connected with the general condition of the animal, the greater the chances of transmission; and when, with a view to breeding, we shall choose only animals having the two-fold character of general vigor of constitution and activity of the mammary system, and place the progeny under favorable circumstances, the qualities will rarely prove defective. Thus far the conclusions of Magne.

Another well-known fact in natural history is, that the size of animals depends very much upon the fertility of the region they inhabit. Where food is abundant and nutritious, they increase in size in proportion to the quantity and quality; and this size, under the same circumstances, will run through generations, unless interrupted by artificial means. So, if the food is more difficult to obtain, and the pastures are short, the pliancy of the animal organization is such that it naturally becomes adapted to it, and the animal is of smaller size; and hence Mr. Cline observes that " the general mistake in crossing has arisen from an attempt to increase the size of a native race of animals, being a fruitless effort to counteract the laws of nature." Mr. Cline also says, in his treatise "On the Form of Animals : " " Experience has proved that crossing has only succeeded in an eminent degree in those instances in which the females were larger than the usual proportion of females to males ; and that it has generally failed when the males were disproportionally large. When the male is much larger than the female, the offspring is generally of an imperfect form ; if the female be proportionally larger than the male, the offspring is generally of an improved form. For instance, if a well-formed large ram be put to ewes proportionally smaller, the lambs will not be so well shaped as their parents ; but, if a small ram be put to larger ewes, the lambs will be of an improved form." " The improvement depends on the principle that the power of the female to supply her offspring with nourishment is in proportion to her size, and to the power of nourishing herself from the excellence of her constitution ; as larger animals. eat more, the larger female may afford most nourishment to her young."

This should, I am inclined to think, be regarded as another principle of breeding,—that, when improvement in form is desired, the size of the female selected should be proportionally larger than the male ; though Lord Spencer, a successful breeder, strongly contested it, and Mr. Dickson, an excellent judge of stock, advised the attempt to build up a new breed by selecting some Zetland cows, a very diminutive breed of Scotch cattle, of good symmetry, points, and handling, and a high-bred West Highland bull to put to them. " The produce would probably be," says he, " a neat, handsome little animal, of a medium size, between the two breeds. The shaggy hide, long horns, symmetry, and fine points, of the West Highlanders, would be imparted to this cross, which would not only be a good feeder and very hardy, but the beef of superior quality. The great point would, of course, be the proper selection of breeding animals. The next step towards improving this would be the crossing of these crosses with a pure Hereford bull,. which would improve the size, and impart still finer points, more substance, with greater aptitude to fatten. By combining these favorite breeds, the produce would, in all probability, be very superior, not only attaining to good weights, but feeding well, and arriving at maturity at an early age. The breeder must not be satisfied and rest here, but go a point further, and cross the heifers of the third cross with a short-horn bull." These successive steps imply the use of a bull of larger breed, though not necessarily, perhaps, pro-portionally larger than the cow, in any individual case.

 This, it will be perceived, is a case of breeding with less reference to the milking or dairy qualities than the grazing. Great milkers are found of all shapes, and the chief object of improving their form is to improve their feeding qualities, or, in other words, to unite, as far as possible, the somewhat incompatible properties of grazing and milking. Graceful, well-rounded, and compact forms, which constitute beauty in the eyes of the grazier, as well as in the estimation of those not accustomed to consider the intrinsic qualities of an animal, or not capable of appreciating them in a milch cow, will very rarely be found united, to any consider able extent, with active mammary glands or milk vessels. The best milkers often look coarse and flabby; for, even if their bony structure is good and symmetrical, they will appear, especially when in milk, to have large, raw bones and sharp points, particularly if they are largely developed in the hind quarters, which is most frequently the case, as is strikingly seen in the form of the Oakes cow, a native animal, the most cele-brated of her time, in Massachusetts, and winner of the first premium at the State Fair of 1816.

She yielded in that year no less than four hundred and sixty-seven and a quarter pounds of butter from May 15th to December 20th, at which time she was giving over eight quarts of milk, beer measure, a day. The weight of her milk in the height of the season, in June, was but forty-four and a half pounds ; not so great as that of some cows of the present day, on far less feed in proportion to their size. Many cows can be named in New England, at the present time, whose yield, under the most favorable circumstances, exceeds fifty pounds a day, and some, whose yield will be fifty-five pounds, on less feed than the Oakes cow had.

Fig. 14. Oakes Cow.

The flesh on the hind quarters of most large milkers bears little proportion to the bone; the hips protrude, the pelvis is broad, the legs far apart, giving great space for the receptacle of large milk-vessels; whilst great flow of blood to the milky glands, incident to this peculiar structure, keeps them in more constant and greater activity than any other organs, so that the muscles develop less than they otherwise would, remain slender, and leave the buttocks and thighs small and narrow. Such animals will seldom acquire the reputation of being beautiful in form, and if they are not decidedly ugly, the owner may console himself with the adage that "handsome is that handsome does."

But, though it is to the influence of the male that we are chiefly to look for improvements in the form, size,

7

muscular development, and general appearance, of our stock, and for transmitting their milking qualities, to a considerable extent, the influence of the female is no less important; and undoubtedly the safest course to pursue, to obtain improved animals, is to select the best-formed animals, on both sides, from the greatest milking families.

With regard to the particular breeds to select for crossing with our natives, opinions will naturally differ widely. Those who are favored with luxuriant pastures and abundance of winter feed will have no objection to large-sized animals, and will naturally wish to obtain or possess grade short-horns. There is no breed in the world to which it is more desirable to resort, under such circumstances, particularly where improvement in form, early maturity, and general symmetry, are sought, in union with other qualities. It is well known that some families of short-horns have been bred for the pail, while most others have been bred chiefly for beef. If resort is had to this breed, therefore, great care and caution should be observed to select bulls from the milking families only; and, unless this is done we shall run the risk of losing the milking qualities of our stock, for which the improvement in form and early maturity can be little compensation, when breeding for the dairy.

It is a remarkable and significant fact that the large dairies of London are nearly filled with the short-horns, or short-horn and Yorkshire grades; and the fact that this breed is selected in such circumstances for the production of milk to supply the milk-market speaks volumes in favor of this cross. It is found that grade short-horns, after yielding extraordinary quantities of milk, during which they very naturally present the most ungainly appearance, will, when dried off and fed, take on flesh very rapidly, and yield large weights of beef. This is one prominent reason for keeping them; and another is, that they occupy less space than would be required to produce the same quantity of milk from smaller animals, which might give even more milk per cow in proportion to size and food consumed.

The cross of the well-bred short-horn and the native or Dutch cows of the dairy districts of New York is very highly esteemed; and six hundred pounds of cheese a year is no uncommon yield for such grades in Herkimer and adjacent counties.

The Ayrshires have been tried in the London dairies, but it was found that they were too difficult to obtain in sufficient numbers, and at sufficiently low prices; and that where quantity was the chief object, as in a milk-dairy, and space a matter of great importance, they could not compete with the short-horn and the Yorkshire cows, and crosses between these races.

It often happens, particularly in milk-dairies, that the farmer is so situated as not to desire to raise his calves, but disposes of them at the highest price to the butcher. He will obtain the greatest weight and the highest quality of veal from the use of a pure-bred short-horn or Hereford bull. But, on poorer pastures, where there is too little feed to bring young stock to their most perfect development, the pure-bred short-horns and high grades of the short-horn are thought, by some, to be too large, and consequently unprofitable. How far this objection to them might be obviated by stall feeding or soiling, and the use of roots, is for each one to consider who has these facilities at command. For most parts of New England they are unquestionably too large to be well maintained.

As to the Herefords, they cannot be recommended for the dairy, either as pure bloods or grades; but in grazing districts, devoted to raising beef or working cattle, they are highly and justly prized.

The same may be said of the North Devons. The pure-bred Devon bull, put to a good, young native cow, produces a beautiful and valuable cross, either for the yoke or the shambles; and if the cow is a remarkably good milker to begin with, and the bull from a milking family, there would be no fear of mate rially lessening the quantity in the offspring, while its form, and other qualities, would probably be greatly improved.

Grade Devons are very much sought for working oxen, and high prices are readily obtained for them, while as beef cattle they are by some highly esteemed. But, unfortunately, very few herds are to be found where attention has been paid to breeding for milk; and great milkers are the exception, and very rarely met with among the pure breeds. In their native country they are bred almost exclusively for beef. The estima- tion in which they are held as dairy stock, even by Devon breeders themselves, both in England and in this country, has been shown in the low value placed upon the development of the udder in the establishment of the scale of points spoken of on a preceding page ; from which it is evident that, in judging of them, it was not contemplated that their milking qualities should be taken into consideration. As cows for the dairy, there- fore, they possess no advantages over our common stock, and we should hardly look for improvement from them in this respect.

The Jerseys, as already seen, are justly celebrated for the richness of their milk and the butter made from it. In this respect no pure breed can excel them. They are, therefore, as a dairy breed, worthy of attention. On farms where the making of butter is an object of pursuit and profit, an infusion of Jersey blood will be likely to secure richness of milk, and high-flavored, delicious butter. Many good judges of stock recommend this cross for dairy purposes; and the chief objection that can be urged against them is that they are, as a breed, very deficient in quantity, which in a milk-dairy would be fatal to them, while, at the same time, they have little to recommend them, as the Devons have, on the score of beautiful forms and symmetrical proportions. Put upon a large and roomy native cow, remarkable as a milker, the produce would be likely to be a very superior animal.

The Ayrshires, as already seen, have been bred with reference both to quality and quantity of milk, and the grades are usually of a very high order. The best milkers I have ever known, in proportion to their size and food, have been grade Ayrshires; and this is also the experience of many who keep dairies for the manufacture of butter and cheese, as well as for the sale of milk. A cross obtained from an Ayrshire bull of good size and a pure-bred short-horn cow will produce a stock which it will be hard to beat at the pail, especially if the cow belong to any of the families of short-horns which have been bred with reference to their milking qualities, as some of them have. I have taken great pains to inquire of dairymen as to the breed or grade of their best cows, and what they consider the best cows for milk for their purposes; and the answer has almost invariably been the Ayrshire and the native. The Ayrshires have by no means been a failure in this country, although I do not think that, as a general thing, we have been so fortunate hitherto as to import the best specimens of them. If any improvement has been made in our dairy stock apart from that effected by a higher and more liberal course of feeding, it has

7*

come, in a great measure, from the Ayrshires; and, had the facilities been offered to cross our common stock with them to greater extent, there can be little doubt that the improvement would have been greater and more perceptible.

It should, however, be said, that in sections where the feed is naturally luxuriant, and adapted to grazing large animals, some families of the short-horns crossed with our natives have produced an equally good stock for cheese and milk dairies.

Before closing this part of the subject, it is proper to observe that among the earlier importations were several varieties of hornless cattle, and that they have been kept distinct in some sections, or where they have been crossed with the common stock there has been a tendency to produce hornless grades. These are not unfrequently known under the name of buffalo cattle. They were, in many cases, supposed to have belonged to the Galloway breed; or, which is more likely, to the Suffolk dun, a variety of the Galloway, and a far better milking stock than the Galloways, from which it sprung. The polled, or hornless cattle, vary in color and qualities, but they are usually very good milkers when well kept, and many of them fatten well, and attain good weights.

The Hungarian cattle have also been imported, to some extent, into different parts of the country, and have been crossed upon the natives with some success. Many other strains of blood from different breeds have contributed to build up the common stock of the country of the present day; and there can be no question that its appearance and value have been largely improved during the last quarter of a century, nor that improvements are still in progress which will lead to satisfactory results in future.

CHAPTER III.

WE have now reviewed the prominent races of cattle found in American dairy herds, and devoted some space to an examination of the principles to be followed in the breeding of dairy stock ; and this has involved, to some extent, the choice of breeds, and the selection of individual animals, with special reference, however, to transmitting and improving their milking properties. But the selection of cows for the dairy is of such importance as to demand the most careful consideration.

The objects of a dairy are three-fold: the production of milk for sale, mainly confined to milk-dairies, and to smaller farms in the vicinity of large towns, where a mixed husbandry is followed ; the production of butter, chiefly confined to farms at a distance from cities and large towns, which furnish a ready market for milk ; and the fabrication of cheese, carried on under circumstances somewhat similar to the manufacture of butter, and sometimes united with it as an object of pursuit, on the farm.

These different objects should, therefore, be kept in view, in the selection of cows ; for animals which would be most profitable for the milk-dairy might be very unprofitable in the butter-dairy — a fact of almost daily experience. The productiveness of the cow does not depend on her breed so much as upon her food and management, her temperament and health, and the activity and energy of the organs of digestion and secretion.

These latter, it is true, depending upon the structure of the chest and other parts, are far better developed, and more permanently fixed, in some races than in others, and are derived more or less by descent, and capable of being transmitted. The breed, therefore, cannot be wholly disregarded, inasmuch as it is an element in forming a judgment of the merits of a milch cow.

Cows, of whatever breed, having the best developed external marks of good milkers, will very rarely disappoint the practised eye or the skilful hand ; while cows of breeds in highest repute for the dairy, and which do not show these marks, will as certainly fail to answer the expectations of those who select them simply for the breed. Those who would obtain skill in judging of these marks, and by means of them be able to estimate the value of a cow, need not expect to attain this end without long study and practical observation, for which some men have far greater talent than others; being able, while examining a particular mark. or favorite characteristic of a milker, to take in all others at a glance, and so, while appearing to form their opinion from one or two important points, actually to estimate the whole development of the animal, while others must examine in detail each point by itself. Long practice is required, therefore, to become an adept in the judgment and selection of milch cows ; but still much assistance may, unquestionably, be derived by careful attention to the external signs which have been long observed to indicate the milking qualities.

It is important, in the first place, to be able to judge of the age of the cow. Few farmers wish to purchase a cow for the dairy after she has passed her prime, which will ordinarily be at the age of nine or ten years, varying, of course, according to care, feeding, &c., in the earlier part of her life.

The most usual mode of forming an estimate of the age of cattle is by an examination of the horn. At three years old, as a general rule, the horns are perfectly smooth; after this, a ring appears near the root, and annually afterward a new one is formed; so that, by adding two years to the first ring, the age is calculated. This is a very uncertain mode of judging. The rings are distinct only in the cow; and it is well known that if a heifer goes to bull when she is two years old, or a little before or after that time, a change takes place in the horn, and the first ring appears; so that a real three-year-old would carry the mark of a four-year-old.

The rings on the horns of a bull are either not seen until five, or they cannot be traced at all; while in the ox they do not appear till he is five years old, and then are often very indistinct. In addition to this, it is by no means an uncommon practice to file the horns, so as to make them smooth, and to give the animal the appearance of being much younger than it really is. This is, therefore, an exceedingly fallacious guide, and we cannot rely on it without being subject to imposition.

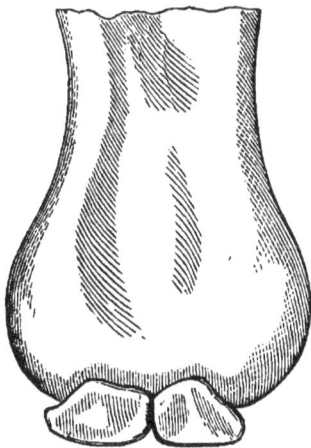

Fig. 15. Teeth at birth.   Fig. 16. Second week.

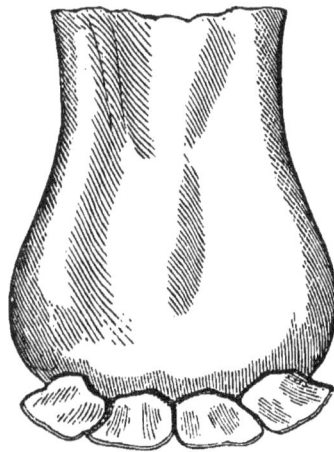

The surest indication of the age is given by the teeth

6

The calf, at birth, will usually have two incisor or front teeth : in some cases just appearing through the gums; in others, fully set, varying as the cow falls short or exceeds her regular time of calving. If she overruns several days, the teeth will have set and attained con- siderable size, as appears in Fig. 15. During the sec- ond week, a tooth will usually be added on each side, and the mouth will generally appear as in Fig. 16; and, before the end of the third week, the animal will gener- ally have six incisor teeth, as shown in Fig. 17; and in a week from that time the full number of incisors will have appeared, as seen in Fig. 18.

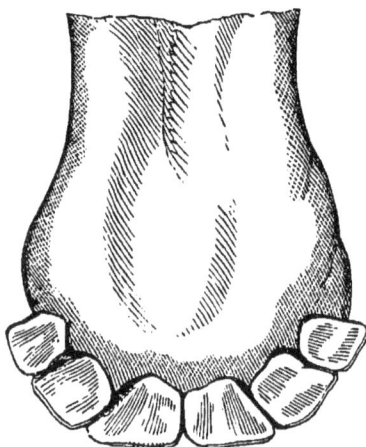

Fig. 17. Third week.   Fig. 18. Month.

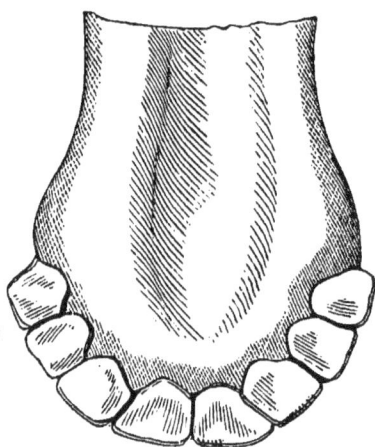

These teeth are temporary, and are often called milk- teeth. Their edge is very sharp; and, as the animal begins to live upon more solid food, this edge becomes worn, showing the bony part of the tooth beneath, and indicates, with considerable precision, the length of time they have been used. The centre or oldest teeth show the marks of age first, and often become some- what worn before the corner teeth appear. At eight weeks, the four inner teeth are nearly as sharp as be-

fore. They appear worn not so much on the outer
edge or line of the tooth, as inside this line ; but, after
this, the edge begins gradually to lose its sharpness,
and to present a more flattened surface ; while the next

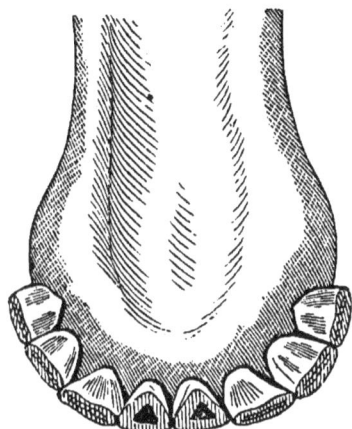

Fig. 19. Five to eight months.   Fig. 20. Ten months.

outer teeth wear down like the four central ones ; and
at three months this wearing off is very apparent, till
at four months all the incisor teeth appear worn, but

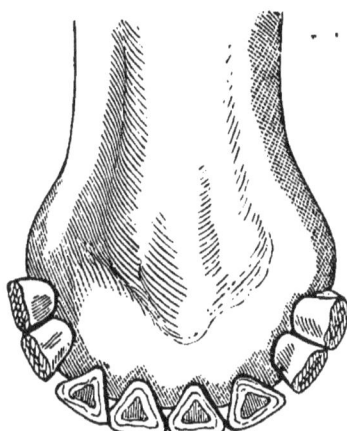

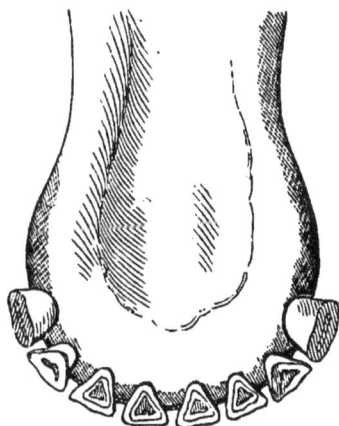

Fig. 21. Twelve months.   Fig. 22. Fifteen months.

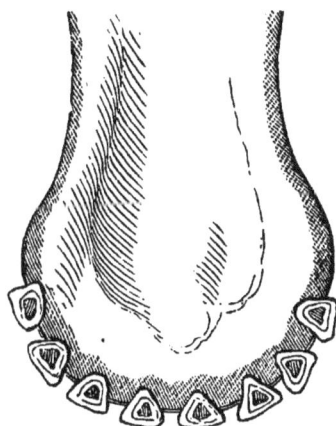

the inner ones the most. Now the teeth begin slowly
to diminish in size by a kind of contraction, as well as

wearing down, and the distance apart becomes more and more apparent.

·From the fifth to the eighth month the inner teeth will usually appear as in Fig. 19; and at ten months this change shows more clearly, as in Fig. 20, and the spaces between them begin to show very plainly, till at a year old they ordinarily present the appearance of Fig. 21; and at the age of fifteen months that shown in Fig. 22, where the corner teeth are not more than half the original size, and the centre ones still smaller.

The permanent teeth are now rapidly growing, and preparing to take the place of the milk-teeth, which are gradually absorbed till they disappear, or are pushed out to give place to the two permanent central incisors, which, at a year and a half, will generally present the appearance indicated in Fig. 23, which shows the internal structure of the lower jaw at this time, with the cells of the teeth, the two central ones protruding into the mouth, the two next pushing up, but not quite

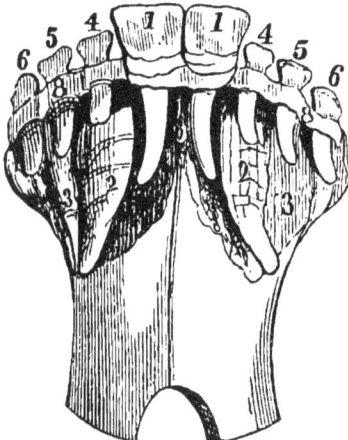

Fig. 23. Eighteen months.   Fig. 24. Two years past

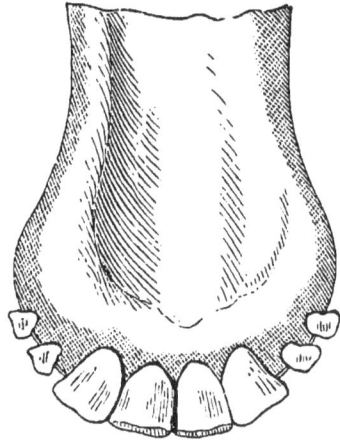

grown to the surface, with the third pair just perceptible. These changes require time; and at two years past the jaw will usually appear as in Fig. 24, where four of the permanent central incisors are seen. After this the other milk-teeth decrease rapidly, but are slow to disappear ; and at three years old the third pair of permanent teeth are but formed, as in Fig. 25 ; and at four years the last pair of incisors will be up, as in Fig. 26 ; but the outside ones are not yet fully grown,

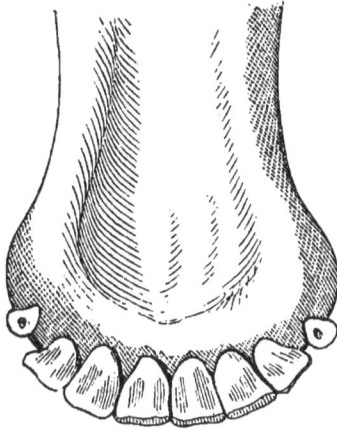

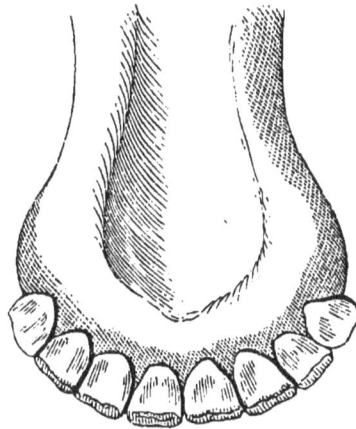

Fig. 25. Three years past.   Fig. 26. Four years past.

and the beast can hardly be said to be full-mouthed till the age of five years. But before this age, or at the age of four years the two inner pairs of permanent teeth are beginning to wear at the edges, as shown in Fig. 26, while at five years old the whole set becomes some-what worn down at the top, and on the two centre ones a darker line appears in the middle, along a line of harder bone, as appears in Fig. 27.

Now will come a year or two, and sometimes three, when the teeth do not so clearly indicate the exact age, and the judgment must be guided by the extent to which the dark middle lines are worn. This will de-pend somewhat upon the exposure and feeding of the animal ; but at seven years these lines extend over all the teeth. At eight years another change begins,

8

which cannot be mistaken. A kind of absorption begins with the two central incisors, slow, at first, but perceptible, and these two teeth become smaller than the rest, while the dark lines are worn into one in all but the

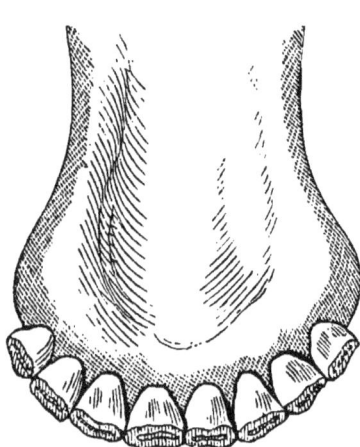

Fig. 27.   Five years past.   Fig. 28.   Ten years past.

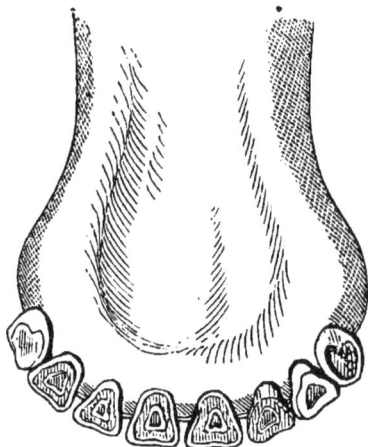

corner teeth, till at ten years four of the central incisors have become smaller in size, with a smaller and fainter mark, as seen in Fig. 28.   At eleven the six inner teeth are smaller than the corner ones; and at twelve all become smaller than they were, while the dark lines are nearly gone, except in the corner teeth, and the inner edge is worn to the gum.

After being satisfied with regard to the age of a cow, we should examine her with reference to her soundness of constitution.   A good constitution is indicated by large lungs, which are found in a deep, broad, and prominent chest, broad and well-spread ribs, a respiration somewhat slow and regular, a good appetite, and if in milk a strong inclination to drink, which a large secretion of milk almost invariably stimulates. · In such cows the digestive organs are active and energetic, and they make an abundance of good blood, which in turn stimulates the activity of the nervous system, and furnishes the milky glands with the means of abundant secretion. Such cows, when dry, readily take on fat. When activity of the milk-glands is found united with close ribs, small and feeble lungs, and a slow appetite, often attended by great thirst, the cow will generally possess only a weak and feeble constitution; and if the milk is plentiful, it will generally be of bad quality, while the animal, if she does not die of diseased lungs, will not take on fat readily when dry and fed.

Other external marks of great milkers have already been given in part. They should be found united, as far as possible; for, though no one of them, however well developed, can be taken as a sure indication of extraordinary milking powers, several of them united may, as a general rule, be implicitly relied on.

In order to have no superfluous flesh, the cow should have a small, clean, and rather long head, tapering towards the muzzle. A cow with a large, coarse head will seldom fatten readily, or give a large quantity of milk. A coarse head increases the proportion of weight of the least valuable parts, while it is a sure indication that the whole bony structure is too heavy. The mouth should be large and broad; the eye bright and sparkling, but of a peculiar placidness of expression, with no indication of wildness, but rather a mild and feminine look. These points will indicate gentleness of disposition. Such cows seem to like to be milked, are fond of being caressed, and often return caresses. The horns should be small, short, tapering, yellowish, and glistening. The neck should be small, thin, and tapering towards the head, but thickening when it approaches the shoulder; the dewlaps small. The fore quarters should be rather small when compared with the hind quarters. The form of the barrel will be large, and each rib should project further than the preceding one, up to the loins. She should be well formed across the hips and in the rump.

The spine or back-bone should be straight and long, rather loosely hung, or open along the middle part, the result of the distance between the dorsal vertebræ, which sometimes causes a slight depression, or sway back. By some good judges this mark is regarded as of great importance, especially when the bones of the hind quarters are also rather loosely put together, leaving the rump of great width, and the pelvis large, and the organs and milk-vessels lodged in the cavities largely developed. The skin over the rump should be loose and flexible. This point is of great importance; and as, when the cow is in low condition, or very poor, it will appear somewhat harder and closer than it otherwise would, some practice and close observation are required to judge well of this mark. The skin, indeed, all over the body, should be soft and mellow to the touch, with soft and glossy hair. The tail, if thick at the setting on, should taper and be fine below.

But the udder is of special importance. It should be large in proportion to the size of the animal, and the skin thin, with soft, loose folds extending well back, capable of great distension when filled, but shrinking to a small compass when entirely empty. It must be free from lumps in every part, and provided with four teats set well apart, and of medium size. Nor are the milk-veins less important to be carefully observed. The principal ones under the belly should be large and prominent, and extend forward to the navel, losing themselves, apparently, in the very best milkers, in a large cavity in the flesh, into which the end of the finger can be inserted; but, when the cow is not in full milk, the milk-vein, at other times very prominent, is not so distinctly

traced; and hence, to judge of its size when the cow is
dry, or nearly so, this vein may be pressed near its end,
or at its entrance into the body, when it will immediately fill up to its full size. This vein does not carry
the milk to the udder, as some suppose, but is the channel by which the blood returns; and its contents consist
of the refuse of the secretion, or what has not been
taken up in forming milk. There are, also, veins in the
udder, and the perineum, or the space above the udder,
and between that and the buttocks, which it is of special importance to observe. These veins should be
largely developed, and irregular or knotted, especially
those of the udder. They may be seen in Figs. 29,
30, 31, &c. They are largest in great milkers.

The knotted veins of the perineum, extending from
above downwards in a winding line, are not readily
seen in young heifers, and are very difficult to find in
poor cows, or cows of only a medium quality. They
are easily found in very good milkers, and, if not
at first apparent, they are made so by pressing upon
them at the base of the perineum, when they swell up,
and send the blood back towards the vulva. They form
a kind of thick network under the skin of the perineum,
raising it up somewhat, in some cases near the vulva,
in others lower down and nearer to the udder. It is
important to look for these veins, as they often form a
very important guide, and by some they would be considered as furnishing the surest indications of the milking qualities of the cow. Their full development almost
always indicates an abundant secretion of milk; but
they are far better developed after the cow has had two
or three calves, when two or three years' milking has
given full activity to the milky glands, and attracted a
large flow of blood. The larger and more prominent
these veins, the better. It is needless to say that in
8*

observing them some regard should be had to the condition of the cow, the thickness of skin and fat by which
they may be surrounded, and the general activity and
food of the animal. Food calculated to stimulate the
greatest flow of milk will naturally increase these veins,
and give them more than usual prominence.

We come now to an examination of the system of
Guénon, whose discovery, whatever may be said of it,
has proved of immense importance to agriculture. Guénon was a man of remarkable practical sagacity, a close
observer of stock, and an excellent judge. This gave
him a great advantage in securing the respect of those
with whom he came in contact, and assisted him vastly
in introducing his ideas to the knowledge of intelligent
men. Born in France, in the vicinity of Bordeaux, in
humble circumstances, he early had the care of cows,
and spent his whole life with them. His discovery, for
which a gold medal was awarded by the agricultural
society of Bordeaux, on the 4th of July, 1837, consisted
in the connection between the milking qualities of the
cow and certain external marks on the udder, and on
the space above it, called the perineum, extending to
the buttocks. To these marks he gave the name of
milk-mirror, or escutcheon, which consists in certain
perceptible spots rising up from the udder in different
directions, forms, and sizes, on which the hair grows
upwards, whilst the hair on other parts of the body
grows downwards. To these spots various names have
l een given, according to their size and position, as tufts,
fringes, figures or escutcheons, which last is the most
common term used. The reduction of these marks into
a system, explaining the value of particular forms and
sizes of the milk-mirror, belongs, so far as I know, exclusively to Guénon, though the connection of the milking qualities of the cow and the size of the ovals with

downward-growing hair on the back part of the udder above the teats was observed and known in Massachu-setts more than forty years ago, and some of the old farmers of that day were accustomed to say that when these spots were large and well developed the cow would be a good milker.

Guénon divided the milk-mirror into eight classes, and each class into eight orders, making in all no less than sixty-four divisions, which he afterwards increased by sub-divisions, making the whole system complicated in the extreme, especially as he professed to be able to judge with accuracy, by means of the milk-mirror, not only of the exact quantity a cow would give, but also the quality of the milk and the length of time it would continue. He tried to prove too much, and the conse-quence was that he was himself frequently at fault, notwithstanding his excellent knowledge of other gene-ral characteristics of milch cows, while others, of less knowledge, and far more liable to err in judgment, were inclined to view the whole system with distrust.

My own attention was called to Guénon's method of judging of cows some eight or ten years ago, and since that time I have examined many hundreds, with a view to ascertain the correctness of its main features, inquiring, at the same time, after the views and opinions of the best breeders and judges of stock, with regard to their experience and judgment of its merits; and the result of my observation has been, that cows with the most perfectly-developed milk-mirrors, or escutcheons, are, with rare exceptions, the best milkers of their breed, and that cows with small and slightly-developed mirrors are, in the majority of cases, bad milkers.

I say the best milkers *of their breed;* for I do not believe that precisely the same sized and formed milk-mirrors on a Hereford or a Devon, and an Ayrshire or a native, will indicate anything like the same or equal milking properties. It will not do, in my opinion, to disregard the general and well-known characteristics of the breed, and rely wholly on the milk-mirror. But I think it may be safely said that, as a general rule the best-marked Hereford will turn out to be the best milker among the Herefords, all of which are poor milkers; the best-marked Devon the best among the Devons, and the best-marked Ayrshire the best among the Ayrshires; that is, it will not do to compare two animals of entirely distinct breeds, by the milk-mirrors alone, without regard to the fixed habits and education, so to speak, of the breed or family to which they belong.

There are cows with very small mirrors, which are, nevertheless, very fair in the yield of milk; and among those with middling quality of mirrors instances of rather more than ordinary milkers often occur, while at the same time it is true that now and then cases occur where the very best marked and developed mirrors are found on very poor milkers. I once owned a cow of most extraordinary marks, the milk-mirror extending out broadly upon the thighs, and rising broad and very distinctly marked to the buttocks, giving every indica-tion, to good judges, of being as great a milker as ever stood over a pail; and yet, when she calved, the calf was feeble and half nourished, and she actually gave too little to feed it. But I believe that this exception, and most others which appear to be direct contradictions, could be clearly explained by the fact, of which I was not aware at the time, that she had been largely over-fed before she came into my possession. I mention this case simply to show how impossible it is to esti-mate with mathematical accuracy either the quantity, the quality, or the duration of the milk, since it is affected by so many chance circumstances, which cannot always be known or estimated by even the most skilful judge; as the food, the treatment, the temperament, accidental diseases, inflammation of the udder, premature calving, the climate and season, the manner in which she has been milked, and a thousand other things which interrupt or influence the flow of milk, without materi- ally changing the size or the shape of the milk-mirror. M. Magne, who appears to me to have simplified and explained the system of Guénon, and to have freed it from many of the useless details with which it is en- cumbered in the original work, while he has preserved all that is of practical value, very justly observes that we often see cows, equally well formed, with precisely the same milk-mirror, and kept in the same circum- stances, yet giving neither equal quantities nor similar qualities of milk. Nor could it be otherwise; for, assuming a particular tuft on two cows to be of equal value at birth, it could not be the same in the course of years, since innumerable circumstances occur to change the activity of the milky glands without chang- ing the form or size of the tuft; or, in other words, the action of the organs depends not merely on their size and form, but, to a great extent, on the general con- dition of each individual.

To give a more distinct idea of the milk-mirror, it will be necessary to refer to the figures, and the explana- tions of these I translate literally from the little work already referred to, the *Choix des Vaches Latières*, or, the Choice of Milch Cows.

The different forms of milk-mirrors are represented by the shaded part of figures 29, 30, 31, etc.; but it is necessary to premise that upon the cows themselves they are always partly concealed by the thighs, the udder, and the folds of the skin, which are not shown, and so they are not always so uniform in nature as they appear in the cuts.

Their size varies as the skin is more or less folded or stretched, while we have supposed in the figures that the skin is uniform or free from folds, but not stretched out. In order to understand the differences which the milk-mirrors present in respect to size, according to the state of the skin, the milk-mirror is shown in two ways in Figs. 52 and 53. In Fig. 53 the proportions are preserved the same as in the other mirrors represented, but an effort is made to represent the folds of the skin ; while in Fig. 52 the mirror is just as it would have been had the folds of the udder been smoothed out, and the skin between the udder and the thighs stretched out ; or, in other words, as if the skin, covered with up growing hair, had been fully extended.

This mirror, but little developed, just as shown in Fig. 53, was observed on a very large Norman cow.

It is usually very easy to distinguish the milk-mirrors by the upward direction of the hair which forms them. They are sometimes marked by a line of bristly hair growing in the opposite direction, which surrounds them, forming a sort of outline by the upward and downward growing hair. Yet, when the hair is very fine and short, mixed with longer hairs, and the skin much folded, and the udder voluminous and pressed by the thighs, it is necessary, in order to distinguish the part enclosed between the udder and the legs, and examine the full size of the mirrors, to observe them attentively, and to place the legs wide apart, and to smooth out the skin, in order to avoid the folds.

The mirrors may also be observed by holding the back of the hand against the perineum, and drawing it from above downwards, when the nails rubbing against the up-growing hair, make the parts covered by it very perceptible.

As the hair of the milk-mirror has not the same direction as the hair which surrounds it, it may often be distinguished by a difference in the shade reflected by it. It is then sufficient to place it properly to the light to see the difference in shade, and to make out the part covered by the upward-growing hair. Most frequently, however, the hair of the milk-mirror is thin and fine, and the color of the skin can easily be seen. If we trust alone to the eye, we shall often be deceived. Thus, in Figs. 52 and 53, the shaded part, which extends from the vulva to the mirror E, represents a strip of hair of a brownish tint, which covered the perineum, and which might easily have been taken for a part of the milk-mirror.

In some countries cattle-dealers shave the back part of the cows. Just after this operation the mirrors can neither be seen nor felt; but this inconvenience ceases in a few days. It may be added that the shaving, designed, as the dealers say, to beautify the cow, is generally intended simply to destroy the milk-mirror, and to deprive buyers of one means of judging of the milking qualities of the cows.

It is not necessary to add that the cows most carefully shaven are those which are badly marked, and that it is prudent to take it for granted that cows so shorn are bad milkers.

Milk-mirrors vary in position, extent, and the figure they represent. They may be divided, according to their position, into mirrors or escutcheons, properly so called, or into lower and upper tufts, or escutcheons. The latter are very small in comparison with the former, and are situated in close proximity to the vulva, as seen at S in Figs. 38, 39, 40, etc. They are very common on cows of bad milking races, but are very rarely seen on the best milch cows. They consist of one or two ovals, or small bands of up-growing hair, and serve to indicate the continuance of the flow of milk. The period is short in proportion as the tufts are large. They must not be confounded with the escutcheon proper, which is often extended up to the vulva. They are separated from it by bands of hair, more or less large, as in Figs. 40, 42, &c.

The mirrors shown in Figs. 38 to 42, and 29 to 35, &c., exist, more or less developed, on nearly all cows, and indicate the quantity of milk, which will be in proportion to their size. Sometimes they form only a small plate on the posterior surface of the udder, as in Fig. 49. In other cases they cover the udder, the inner surface of the legs and the thighs, the perineum, and a part of the buttocks, as in Figs. 29, 30, 31, &c.

Two parts may be distinguished in the lower tufts : one situated on the udder, the legs, and the thighs, as at M M, Fig. 30 ; and the other on the perineum, extending sometimes more or less out upon the thighs, as at P P, in the same figure.

The first part is represented by itself, in Figs. 37 and 49. We shall call the former *mammary*, and the latter *perinean*. The former is sometimes large, extending over the milky glands, the thighs, and the legs, as shown in Figs. 29 to 37; and sometimes circumscribed, or more or less checked over with tufts of downward-growing hair, as in Figs. 43 to 52. It is sometimes terminated towards the upper part of the udder by a horizontal line, straight, as in Fig. 37, or angular, as in Fig. 49 ; but more frequently it continues without interruption over the perineum, and constitutes the *perinean* part.

This presents a large band, Fig. 30, straight, as in Fig. 43, and bounded on the sides by two parallel lines,

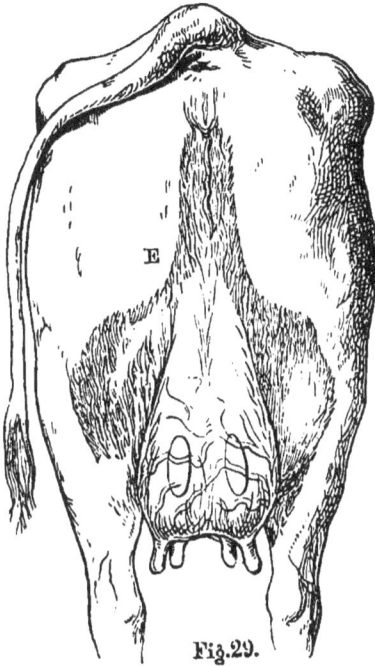

Fig.29.

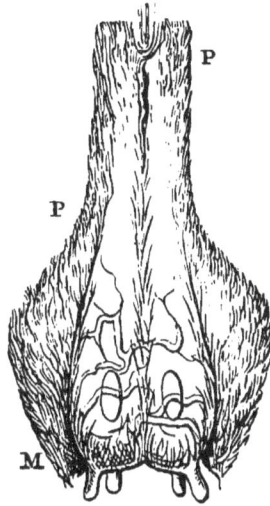

Fig.30.

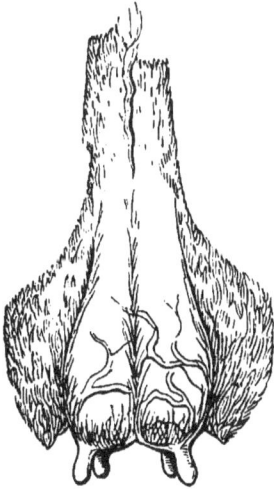

Fig.31.

9

Fig.32.

7

as seen in the same figures, or by curved lines, as in Fig. 34. It sometimes rises scarcely a fourth part up the perineum, as in Fig. 38 ; at others, it reaches or passes beyond that part, forming a straight band, as in Figs. 35 and 43, or is folded into squares, as in Figs. 31 and 36, or truncated, Fig. 38, or terminated by one or several points, Figs. 32, 33, 41, 50. In some cows this band extends as far as the base of the vulva, Figs. 40 and 48 ; in others, it embraces more or less of the lower part of the vulva, Figs. 29, 30, 39, and 47.

Milk-mirrors are sometimes symmetrical, as in Figs. 29, 30, 34, 35, 37, and 38 ; sometimes without symmetry, as in Figs. 42, 45, and 50. When there is a great difference in the extent of the two halves, it almost always happens that the teats on the side where the mirror is best developed give, as we shall see, more milk than those of the opposite side. We will remark here that the left half of the mirror is almost always the largest ; and so, when the perinean part is folded into a square, it is on this side of the body that it unfolds, as in Figs. 31, 36, and 42. Of three thousand cows in Denmark, M. Andersen found only a single one whose escutcheon varied even a little from this rule. We have observed the contrary only in a single case, and that was on a bull. The perinean part of the mirror formed a band of an inch to an inch and a half in breadth, irregular, but situated, in great measure, on the right side of the body. Stretching towards the upper part of the perineum, it formed a kind of square, with a small projecting point on the right, Fig. 51.

The mirrors having a value in proportion to the space they occupy, it is of great importance to attend to all the rows of down-growing hairs, which diminish its extent of surface, whether these tufts are

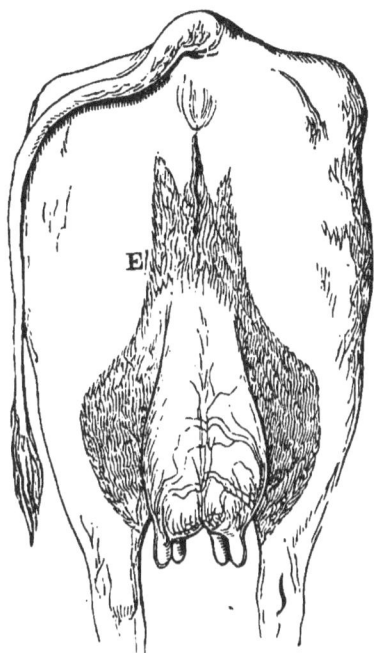

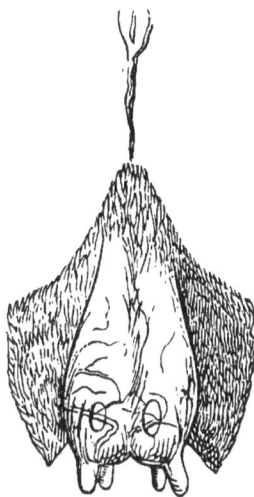

Fig. 34.

Fig. 33.

Fig. 36.

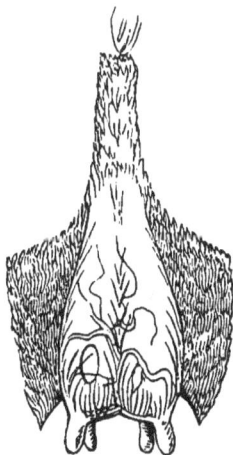

Fig. 35.

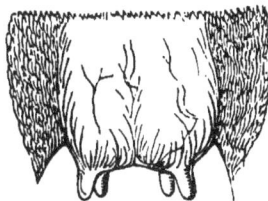

Fig. 37.

in the midst of the mirror, Figs. 45, 46, and 47, or form indentations on its edges, as in Figs. 42, 44, 45, 46, and 48.

These indentations, concealed in part by the folds of the skin, are sometimes seen with difficulty; but it is important to take them into account, since in a great many cows they materially lessen the size of the mir-ror. We often find cows whose milk-mirror at first sight appears very large, but which are only medium milkers; and it will usually be found that lateral indent-ations greatly diminish the surface of up-growing hair. Many errors are committed in estimating the value of such cows, from a want of attention to the real extent of the milk-mirror.

All the interruptions in the surface of the mirror indicate a diminution of the quantity of milk, with the exception, however, of small oval or elliptical plates which are found in the mirror, on the back part of the udders of the best cows, as in Figs. 29, 30, 32, 34, 35, 36, and 40. These ovals have a peculiar tint, which is occasioned by the downward direction of the hair which forms them. In the best cows these ovals exist with the lower mirrors very well developed, as in Figs. 29, 30, and 32.

In fine, we should state that in order to determine the extent and significance of a mirror it is necessary to consider the state of the perineum as to fat, and of the fulness of the udder. In a fat cow, with an in-flated udder, the mirror would appear larger than it really is; whilst in a lean cow, with a loose and wrinkled udder, it appears smaller. Fat will cover faults, a fact to be kept in mind in selecting a cow.

In bulls, Fig. 51, the mirrors present the same pecu-liarities as in cows; but they are less varied in their form, and especially much less in size. This will easily

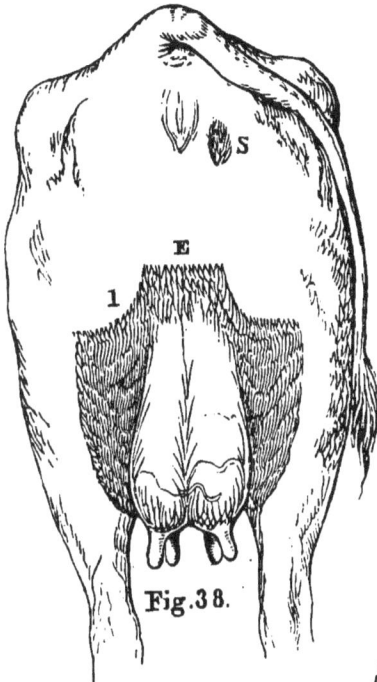

Fig. 38.

Fig. 39.

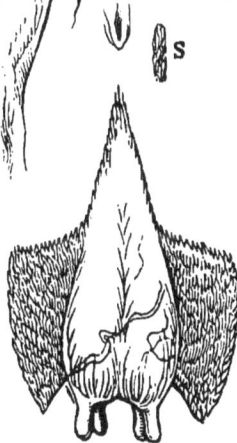

Fig. 41.

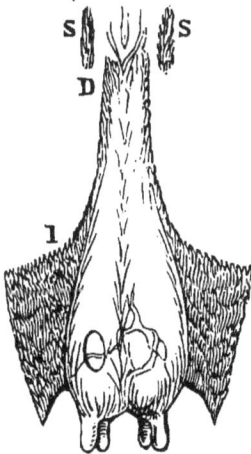

Fig. 40.

Fig. 42.

9*.

be understood from the explanation of mirrors given on a preceding page.

In calves the mirrors show the shapes they are after-wards to have, only they are more contracted, because the parts which they cover are but slightly developed. They are easily seen after birth; but the hair which then covers them is long, coarse, and stiff; and when this hair falls off, the calf's mirror will resemble that of the cow, but be of less size.

With calves, however, it should be stated, in addition to what has already been said, that the milk-mirrors are more distinctly recognized on those from cows that are well kept, and that they will generally be fully devel-oped at two years old. Some changes take place in the course of years, but the outlines of the mirror appear prominent at the time of advanced pregnancy, or, in the case of cows giving milk, at the times when the udder is more distended with milk than at others.

The classification adopted by Magne appears still further to simplify the whole method, and to, bring it within the easy reach and comprehension of every one who will examine the figures and the explanations con-nected with them. He divides cows, according to the quantity they give, into four classes : First, the very good ; second, the good; third, the medium ; and fourth, the bad.

In the first class he places cows both parts of whose milk-mirror, the mammary and the perinean, are large, continuous, uniform, covering at least a great part of the perineum, the udder, the inner surface of the thighs, and extending more or less out upon the legs, as in Figs. 29 to 33, with no interruptions, or, if any small ones, oval in form, and situated on the posterior face of the udder, Figs. 29, 30, and 32.

Such mirrors are found on most very good cows

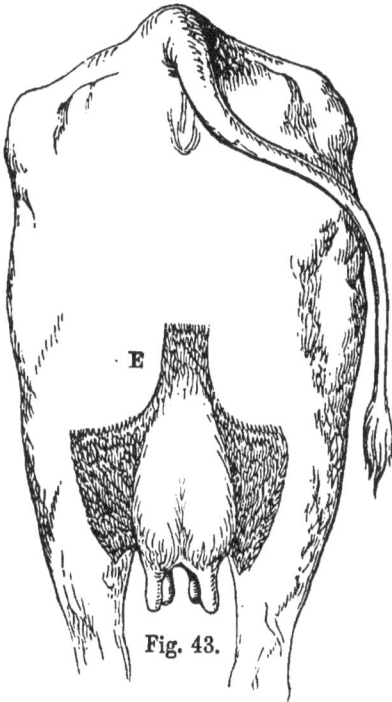

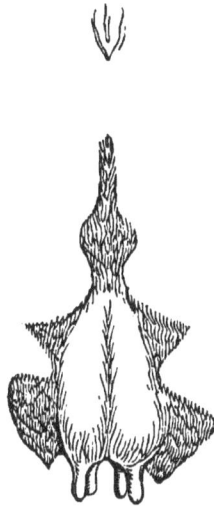

Fig. 44.

Fig. 43.

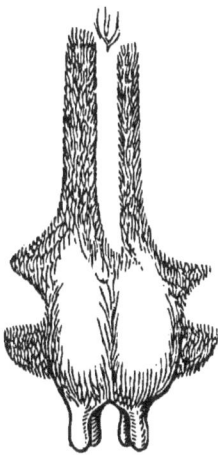

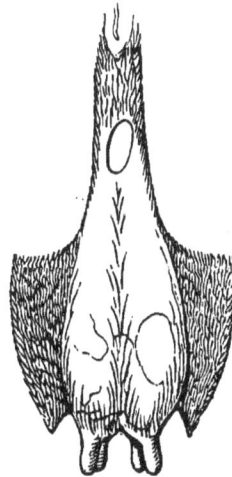

Fig. 45.   Fig. 46.   Fig. 47.

but may also be found on cows which can scarcely be called good, and which should be ranked in the next class. But cows, whether having very well-developed mirrors or not, may be reckoned as very good, and as giving as much milk as is to be expected from their size, feed, and the hygienic circumstances in which they are kept, if they present the following characteristics:

Veins of the perineum large, as if swollen, and visible on the exterior, as in Figs. 29—32, or which can be easily made to appear by pressing upon the base of the perineum; veins of the udder large and knotted, milk-veins large, often double, equal on both sides, and forming zig-zags under the belly.

To the signs furnished by the veins and by the mirror may be added also the following marks: A uniform, very large and yielding udder, shrinking much in milking, and covered with soft skin and fine hair; good constitution, full chest, regular appetite, and great propensity to drink. Cows rather inclining to be poor than fat. Soft, yielding skin, short, fine hair, small head, fine horns, bright, sparkling eye, mild expression, feminine look, with a fine neck.

Cows of this first class are very rare. They give, even when small in size, from ten to fourteen quarts of milk a day, and the largest sized from eighteen to twenty-six quarts a day, and even more. Just after calving, if arrived at maturity and fed with good, wholesome, moist food in sufficient quantity and quality, adapted to promote the secretion of milk, they can give about a pint of milk for every ten ounces of hay, or its equivalent, which they eat.

They continue in milk for a long period. The best never go dry, and may be milked even up to the time of calving, giving from eight to twelve quarts of milk a day. The Dutch cow, Fig. 54, was giving daily

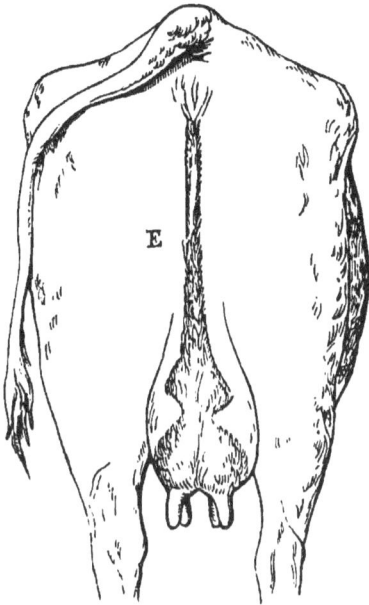

Fig.49.

Fig.48.

Fig.50.

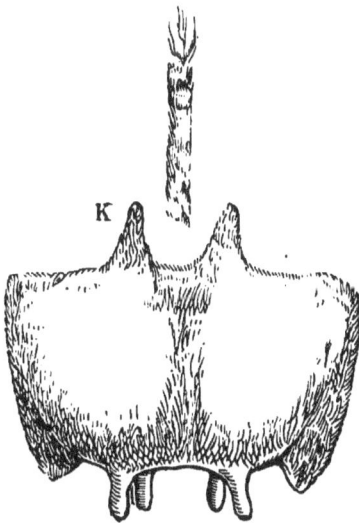

Fig.51.

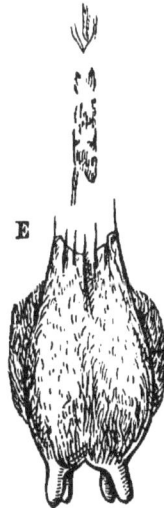

Fig.52.

Fig.53.

twenty-two quarts of milk, a year after calving. But even the best cows often fall short of the quantity of milk they are able to give, from being fed on food that is too dry, or not sufficiently varied, or not rich enough in nutritive qualities, or deficient in quantity.

The second class is that of *good cows*; and to this belong the best commonly found in the market and among the cow-feeders of cities.

They have the mammary part of the milk-mirror well developed, but the perinean part contracted or wholly wanting, as in Figs. 34 and 37; or both parts of the mirror are moderately developed, or slightly indented, as Figs. 35 and 36. Figs. 38, 39, 40, and 41, belong also to this class, in the lower part; but they denote cows which, as the upper mirrors, s s s, indi- cate, dry up sooner when again in calf.

These marks, though often seen on many good cows, should be considered as certain only when the veins of the perineum form, under the skin, a kind of network, which, without being very apparent, may be felt by a pressure on them; when the milk-veins on the belly are well developed, though less knotted and less prominent than in cows of the first class; in fine, when the udder is well developed, and presents veins which are suffi- ciently numerous, though not very large.

It is necessary, then, as in the preceding class, to have a mistrust of cows in which the mirror is not accompanied by large veins. This remark applies especially to cows which have had several calves, and are in full milk. They are medium or bad, let the milk- mirror be what it may, if the veins of the belly are not large, and those of the udder apparent.

The general characteristics which depend on form and constitution combine less than in cows of the pre-

Fig. 54. A Good Milch Cow.

ceding class the marks of good health and excellent constitution with those of a gentle and feminine look.

Small cows of this class give from seven to ten or eleven quarts of milk a day, and the largest from thirteen to seventeen quarts. They can be made to give three fourths of a pint of milk, just after calving, for every ten ounces of hay consumed, if well cared for, and fed in a manner favorable to the secretion of milk.

They hold out long in milk when they have no upper mirrors or tufts. At seven or eight months in calf, they may give from five to eight quarts a day.

The third class consists of *middling cows*. When the milk-mirror really presents only the lower or mammary part slightly developed or indented, and the perinean part contracted, narrow, and irregular, as in Figs. 42 to 47, the cows are middling. The udder is slightly developed or hard, and shrinks very little after milking. The veins of the perineum are not apparent, and those which run along the lower sides of the abdomen are small, straight, and sometimes unequal. In this case the mirror is not symmetrical, and the cow gives more milk on the side where the vein is largest.

These cows often have large heads, and a thick and hard skin. Being ordinarily in good condition, and even fat, they are beautiful to look at, and seem to be well formed. Many of them are nervous and restive, and not easily approached.

Cows of this class give, according to size, from three or four to ten quarts of milk. They very rarely give, even in the most favorable circumstances, half a pint for every ten ounces of hay which they consume.

The milk diminishes rapidly, and dries up wholly the fourth or fifth month in calf.

The fourth class is composed of *bad cows*. As they are ordinarily in good condition, these cows are often the most beautiful of the herd and in the markets. They have fleshy thighs, thick and hard skin, a large and coarse neck and head, and horns large at the base.

The udder is hard, small, and fleshy, with a skin covered with long, rough hair. No veins are to be seen either on the perineum or the udder, while those of the belly are very slightly developed, and the mirrors are ordinarily small, as in Figs. 48, 49, and 50.

With these characteristics, cows give only a few quarts of milk a day, and dry up a short time after calving. Some such can scarcely nourish their calves, even when they are well cared for and well fed.

Sickly habits, chronic affections of the digestive organs, the chest, the womb, and the lacteal system, sometimes greatly affect the milk secretions, and cause cows troubled with them to fall from the first or second to the third, and sometimes to the fourth class.

The above classification is very similar to that of Pabst, a German farmer of large experience and observation of stock, who, with a view to simplify the method of Guénon, and render it of greater practical value to the farmer, made five divisions or classes, consisting of, 1st, *Very good* or *extraordinary;* 2d, *Good* or *good middling;* 3d, *Middling* and *little below middling;* 4th, *Small;* and, 5th, *Very bad milkers.*

These classifications, adopted by Magne, Pabst, and other good breeders and judges of cows, appear to me to be far more simple and satisfactory than the more extended and complicated classification of Guénon himself. Without pretending to be able to judge with any accuracy of the quantity, the quality, or the duration, which any particular size or form of the mirror will indicate, they give to Guénon the full credit of his important discovery of the escutcheon, or milk-mirror, as a new and very valuable element in forming our judgment of the milking qualities of a cow ; and simply assert, with respect to the duration or continuance of the flow of milk, that the mirror that indicates the greatest quantity will also indicate the longest duration. The mirror- forms, in other words, an important additional mark or point for distinguishing good milk-

10

ers; and it is safe to lay it down as a rule that, in the selection of milch cows, as well as in the choice of young animals as breeders, we should, by all means, examine and consider the milk-mirror, but not limit or confine ourselves exclusively to it, and that other and long-known marks should be equally regarded.

But there are cases where a knowledge and careful examination of the form and size of the mirror becomes of the greatest importance. It is well known that certain signs or marks of great milkers are developed only as the capacities of the animal herself are fully and completely developed by age. The milk-veins, for instance, are never so large and prominent in heifers and young cows as in old ones, and the same may be said of the udder, and the veins of the udder and perineum; all of which it is of great importance to observe in the selection of milch cows. Those signs, then, which in cows arrived at maturity are almost sufficient in themselves to warrant a conclusion as to their merits as milkers, are, to a great extent, wanting in younger animals, and altogether in calves, of which there is often doubt whether they shall be raised; and here a knowledge of the form of the mirror is of immense advantage, since it gives, at the outset, and before any expense is incurred, a somewhat reliable means of judging of the future milking capacities of the animal or, if a male, of the probability of his transmitting milking qualities to his offspring.

It will be seen, from an examination of the points of a good milch cow, that, though the same marks which indicate the greatest milking qualities may not indicate any great aptitude to fatten, yet that the signs which indicate good fattening qualities are included among the signs favorable to the production of milk, such as soundness of constitution, indicated by good organs of digestion and respiration, fineness and mellowness of the skin and hair, quietness of disposition, which inclines the animal to rest and lie down in chewing the cud, and other marks which are relied on by graziers in selecting animals to fatten.

In buying dairy stock the farmer generally finds it for his interest to select young heifers. They give the promise of longer usefulness. But it is often the case that older cows are selected with the design of using them for the dairy for a limited period, and then feeding them for the butcher. In either case, it is advisable, as a rule, to choose animals in low or medium condition. The farmer cannot ordinarily afford to buy fat; it is more properly his business to make it, and to have it to sell. Good and well-marked cows in poor condition will rapidly gain in flesh and products when removed to better pastures and higher keeping, and they cost less in the original purchase.

It is unnecessary to say that regard should be had to the quality of the pasturage and keeping which a cow has previously had, as compared with that to which she is to be subjected. The size of the animal should also be considered with reference to the fertility of the pastures into which she is to be put. Small or medium-sized animals accommodate themselves to ordinary pastures far better than large ones. Where a very large cow will do well, two small ones will usually do better: while the large animal might fail entirely where two small ones would do well. It is better to have the whole herd, so far as may be, uniform in size; for, if they vary greatly, some may get more than they need, and others will not have enough. This, however, can not always be brought about.

Fig. 56. A Good Dairy Cow

CHAPTER IV.

No branch of dairy farming can compare in importance with the management of cows. The highest
success will depend very much upon it, whatever breed
be selected, and whatever amount of care and attention
be given to the points of the animals ; for experience
will show that very little milk comes out of the bag
that is not first put into the throat. It is poor economy, therefore, to attempt to keep too many cows for
the amount of feed we have ; for it will generally be
found that one good cow well bred and well fed will
yield as much as two ordinary cows kept in the ordinary way, while a saving is effected both in labor and
room required, and in the risks on the capital invested.
If the larger number on poorer feed is urged for the
sake of the manure, which is the only ground on which
it can be put, it is sufficient to remark that it is a very
expensive way of making manure. It is not too much
to say that a proper regard to profit and economy
would require many an American farmer to sell off
nearly half his cows, and to feed the whole of his hay
and roots hitherto used into the remainder.

A certain German farmer was visited, one day, by
some Swiss from over the border, who desired to buy
of him all the milk of his cows for the purpose of
making cheese. Not being able to agree upon the

10*   8

terms, he finally proposed to let them take the entire
charge of his cows, and agreed to furnish feed amply
sufficient, the Swiss assuming the whole care of feeding
it out, and paying a fixed price by measure for all the
milk. "I found myself, at once," says he, "under the
necessity of selling almost half my cows, because the
Swiss required nearly double the quantity of fodder
which the cows had previously had, and I was well satisfied that all the produce I could raise on my farm
would be far from sufficient to feed in that way the
number of cows I had kept. I was in despair at finding them using such a quantity of the best quality of
feed, though it was according to the strict letter of the
contract, especially as I knew that I had given my cows
rather more than the quantity of food recommended by
men in whom ı had perfect confidence. Thus, while
Thaër names twenty-three pounds of hay, or its equivalent, as food sufficient for a good-sized cow, I gave mine
full twenty-seven pounds. But, if the change effected
in the management of my cows was great, the result
was still more striking. The quantity of milk kept
increasing, and it reached the highest point when the
cows attained the condition of the fat kine of Pharaoh's
dream. The quantity of milk became double, triple,
and even quadruple, what it had been before; so that,
if I should compare the product with that previously
obtained, a hundred pounds of hay produced three
times more milk than it had produced with my old
mode of feeding. Such results, of course, attracted my
attention to this branch of my farming. It became a
matter of pleasure; and my observations were followed
up with great care, and during several years I devoted
a large part of my time to it. I even went so far as to
procure scales for weighing the food and the animals, in
order to establish exact data on the most positive basis."

The conclusions to which he arrived were, that an animal, to be fully fed and satisfied, requires a quantity of food in proportion to its live weight; that no feed could be complete that did not contain a sufficient amount of nutritive elements; hay, for example, being more nutritive than straw, and grains than roots. He found, too, that the food must possess a bulk sufficient to fill up to a certain degree the organs of digestion or the stomach; and that, to receive the full benefit of its food, the animal must be wholly satisfied, as, if the stomach is not sufficiently distended, the food cannot be properly digested, and of course many of the nutritive principles it contains would not be perfectly assimilated. An animal regularly fed eats till it is satisfied, and no more than is requisite. A part of the nutritive elements in hay and other forage-plants is needed to keep an animal on its feet, — that is, to keep up its condition, — and if the nutrition of its food is not sufficient for this the weight decreases, and if it is more than sufficient the weight increases, or else this excess is consumed in the production of milk or in labor. About one sixtieth of their live weight in hay, or its equivalent, will keep horned cattle on their feet; but, in order to be completely nourished, they require about one thirtieth in dry substances, and four thirtieths in water, or other liquid contained in their food. The excess of nutritive food over and above what is required to sustain life will go in milch cows generally to the production of milk, or to the growth of the fœtus, but not in all cows to an equal extent; the tendency to the secretion of milk being far more developed in some than in others.

With regard to the consumption of food in proportion to the live weight of the animal, however far it may apply as a general principle, it should, I think, be taken with some qualifications The proportion is probably not uniform as applied to all breeds indiscriminately, though it may be more so as applied to animals of the same breed. Bakewell's idea was that the quantity of food required depended much on the shape of the barrel; and it is well known that an animal of a close, compact, well-rounded barrel will consume less than one of an opposite make.

The variations in the yield of milch cows are caused more by the variations in the nutritive elements of their food than by a change of the form in which it is given. "A cow, kept through the winter on mere straw," says a practical writer on this subject, "will cease to give milk; and, when fed in spring on green forage, will give a fair quantity of milk. But she owes the cessation and restoration of the secretion to respectively the diminution and the increase of her nourishment, and not at all to the change of form, or of outward substance, in which the nourishment is administered. Let cows receive through winter nearly as large a proportion of nutritive matter as is contained in the clover, lucerne, and fresh grasses, which they eat in summer, and, no matter in what precise substance or mixture that matter may be contained, they will yield a winter's produce of milk quite as rich in caseine and butyraceous ingredients as the summer's produce, and far more ample in quantity than almost any dairyman with old-fashioned notions would imagine to be possible. The great practical error on this subject consists not in giving wrong kinds of food, but in not so proportioning and preparing it as to render an average ration of it equally rich in the elements of nutrition, and especially in nitrogenous elements, as an average ration of the green and succulent food of summer."

We keep too much stock for the quantity of good

and nutritious food which we have for it; and the consequence is cows are, in nine cases out of ten, poorly
wintered, and come out in the spring weakened, if not,
indeed, positively diseased, and a long time is required
to bring them into a condition to yield a generous
quantity of milk.

It is a hard struggle for a cow reduced in flesh and
in blood to fill up the wasted system with the food
which would otherwise have gone to the secretion of
milk; but, if she is well fed, well housed, well littered,
and well supplied with pure, fresh water, and with
roots, or other *moist* food, and properly treated to the
luxury of a frequent carding, and constant kindness,
she comes out ready to commence the manufacture of
milk under favorable circumstances.

Keep the cows constantly in good condition, ought,
therefore, to be the motto of every dairy farmer, posted
up over the barn-door, and over the stalls, and over the
milk-room, and repeated to the boys whenever there is
danger of forgetting it. It is the great secret of success, and the difference between success and failure
turns upon it. Cows in milk require more food in proportion to their size and weight than either oxen or
young cattle.

In order to keep cows in milk well and economically,
regularity is next in importance to a full supply of
wholesome and nutritious food. The healthy animal
stomach is a very nice chronometer, and it is of the
utmost importance to observe regular hours in feeding,
cleaning, and milking. This is a point, also, in which
very many farmers are at fault — feeding whenever it
happens to be convenient. The cattle are thus kept in
a restless condition, constantly expecting food when
the keeper enters the barn, while, if regular hours are
strictly adhered to, they know exactly when they are

to be fed, and they rest quietly till the time arrives. Go into a well-regulated dairy establishment an hour before the time of feeding, and scarcely an animal will rise to its feet; while, if it happens to be the hour of feeding, the whole herd will be likely to rise and seize their food with an avidity and relish not to be mistaken.

With respect to the exact routine to be pursued, no rule could be prescribed which would apply to all cases; and each individual must be governed much by circumstances, both in respect to the particular kinds of feed at different seasons of the year, and the system of feeding. I have found in my own practice, and in the practice of the most successful dairymen, that, in order to encourage the largest secretion of milk in stalled cows, one of the best courses is, to feed in the morning, either at the time of milking — which I prefer — or immediately after, with cut feed, consisting of hay, oats, millet, or corn-stalks, mixed with shorts, and Indian, linseed, or cotton-seed meal, thoroughly moistened with water. If in winter, hot or warm water is far better than cold. If given at milking-time, the cows will generally give down the milk more readily. The stalls and mangers ought always to be well cleaned out first.

Roots and long hay may be given during the day; and at the evening milking, or directly after, another generous meal of cut feed, well moistened and mixed, as in the morning. No very concentrated food, like grains alone or oil-cakes, should, it seems to me, be fed early in the morning on an empty stomach, though it is sanctioned by the practice in the London milk-dairies. The processes of digestion go on best when the stomach is sufficiently distended; and for this purpose the bulk of food is almost as important as the nutritive qualities. The flavor of some roots, as cabbages and turnips, is more apt to be imparted to the flesh and milk when fed on an empty stomach than otherwise. After the cows have been milked, and have finished their cut feed, they are carded and curried down, in well-managed dairies, and then either watered in the stall, which in very cold or stormy weather is far preferable, or turned out to water in the yard. When they are out, if they are let out at all, the stables are put in order; and, after tying them up, they are fed with long hay, and left to themselves till the time of next feeding. This may consist of roots, such as cabbages, beets, carrots, or turnips, sliced, or of potatoes, a peck, or, if the cows are very large, a half-bushel each, and cut feed again at the evening milking, as in the morning, after which water in the stall, if possible.

The less cows are exposed to the cold of winter, the better. They eat less, thrive better, and give more milk, when kept housed all the time, than when exposed to the cold. Caird mentions a case where a herd of cows, which had been usually supplied from troughs and pipes in the stalls, were, on account of an obstruction in the pipes, obliged to be turned out twice a day to be watered in the yard. The quantity of milk instantly decreased, and in three days the falling off became very considerable. After the pipes were mended, and the cows again watered as before, in their stalls, the flow of milk returned. This, however, will be governed much by the weather; for in very mild, warm days it may be judicious not only to let them out, but to allow them to remain out for a short time, to exercise.

Any one can arrange the hour for the several processes named above, to suit himself; but, when once fixed, let it be rigidly and regularly followed. If the regular and full feeding be neglected for even a day, the yield of milk will immediately decline, and it will be very difficult to restore it. It may safely be asserted, as the result of many trials and long practice, that a larger flow of milk follows a complete system of regularity in this respect than from a higher feeding where this system is not adhered to.

One prime object which the dairyman should keep constantly in view is, to maintain the animal in a sound and healthy condition. Without this, no profit can be expected from a milch cow for any considerable length of time; and, with a view to this, there should be an occasional change of food. But, in making changes, great care is required to supply an equal amount of nourishment, or the cow falls off in flesh, and eventually in milk. We should therefore bear in mind that the food consumed goes not alone to the secretion of milk, but also to the growth and maintenance of the bony structure, the flesh, the blood, the fat, the skin, and the hair, and in exhalations from the body. These parts of the. body consist of different organic constituents. Some are rich in nitrogen, as the fibrin of the blood, albumen, &c.; others destitute of it, as fat ; some abound in inorganic salts, phosphate of lime, salts of potash, &c. To explain how the constant waste of these substances may be supplied, Dr. Voelcker observes that the albumen, gluten, caseine, and other nitrogenized principles of food, supply the animal with materials required for the formation of muscle and cartilage ; they are, therefore, called flesh-forming principles. ·

"Fats, or oily matters of the food," says he, "are used to lay on fat, or for the purpose of sustaining respiration.

"Starch, sugar, gum, and a few other non-nitrogenized substances, consisting of carbon, oxygen, and hydrogen, supply the carbon given off in respiration, or they are used for the production of fat.

"Phosphates of lime and magnesia in food principally

furnish the animal with the materials of which the bony skeleton of its body consists.

"Saline substances — chlorides of sodium and potassium, sulphate and phosphate of potash and soda, and some other mineral matters occurring in food — supply the blood, juice of flesh, and various animal juices, with the necessary mineral constituents.

"The healthy state of an animal can thus only be preserved by a mixed food; that is, food which contains all the proximate principles just noticed. Starch or sugar alone cannot sustain the animal body, because neither of them furnishes the materials to build up the fleshy parts of the animal. When fed on substances in which an insufficient quantity of phosphates occurs, the animal will become weak, because it does not find any bone-producing principles in its food. Due attention, therefore, ought to be paid by the feeder to the selection of food which contains all the kinds of matter required, nitrogenized as well as non-nitrogenized, and mineral substances; and these should be mixed together in the proportion which experience points out as best for the different kinds of animals, or the particular purpose for which they are kept."

"On the nutrition of cows for dairy purposes," Dr. Voelcker still further observes that "milk may be regarded as a material for the manufacture of butter or of cheese ; and, according to the purpose for which the milk is intended to be employed, whether for the manufacture of butter or the production of cheese, the cow should be differently fed.

"Butter contains carbon, hydrogen, and oxygen, and no nitrogen. Cheese, on the contrary, is rich in nitrogen. Food which contains much fatty matter, or substances which in the animal system are readily converted into fat, will tend to increase the proportion of

11

cream in milk. On the other hand, the proportion of caseine or cheesy matter in milk is increased by the use of highly nitrogenized food. Those, therefore, who desire much cream, or who produce milk for the manufacture of butter, select food likely to increase the proportion of butter in the milk. On the contrary, where the principal object is the production of milk rich in curd,— that is, where cheese is the object of the farmer, —clover, peas, and bean-meal, and other plants which abound in legumine,— a nitrogenized organic compound, almost identical in properties and composition with caseine, or the substance which forms the curd of milk,— will be selected." And so the quality, as well as the quantity, of butter in the milk, depends on the kind of food consumed, and on the general health of the animal. Cows fed on turnips in the stall always produce butter inferior to that of cows living upon the fresh and aromatic grasses of the pastures.

Succulent food in which water abounds — the green grass of irrigated meadows, green clover, brewers' refuse, distillers' refuse, etc. — increases the quantity, rather than the quality, of the milk; and by feeding these substances the milk-dairyman studies his own interest, and makes thin milk, without diluting it with water, though, in the opinion of some, this may be no more legitimate than watering the milk.

But, though the yield of milk may be increased by succulent or watery food, it should be given so as not to interfere with the health of the cow.

Food rich in starch, gum, or sugar, which are the respiratory elements, an excess of which goes to the production of fatty matters, increases the butter in milk. Quietness promotes the secretion of fat in animals and increases the butter. Cheese will be increased by food rich in albumen, such as the leguminous plants.

The most natural, and of course the healthiest food for milch cows in summer, is the green grass of the pastures; and when these fail from drought, or over-stocking, the complement of nourishment may be made up with green clover, green oats, barley, millet, or corn-fodder, and cabbage-leaves, or other succulent vegetables; and if these are wanting, their place may be partly supplied with shorts, Indian-meal, linseed or cotton-seed meal. Green grass is more nutritious than hay, which always loses more or less of its nutritive qualities in curing; the amount of the loss depending chiefly on the mode of curing, and the length of exposure to sun and rain. But, apart from this, grass is more easily and completely digested than hay, though the digestion of hay may be greatly aided by cutting and moistening, or steaming; and by this means it is rendered more readily available, and hence far better adapted to promote a large secretion of milk — a fact too often overlooked by many even intelligent farmers.

That green grass is better adapted than most other kinds of food to promote a large flow of milk, may be be seen from the following table, from which it will appear that greater attention should be given to the proper constituents of food for milch cows. Two cows were taken in the experiment.

Food of two cows.	Milk in five days.	Butter in five days.	Nitrogen in food in five days.
1. Grass, ...	114 lbs.	3.50 lbs.	2.32 lbs
2. Barley and hay, . . .	107	3.43	3.89
3. Malt and hay, . . .	102	3.20	3.34
4. Barley, molasses, and hay, ...	106	3.44	3.82
5. Barley, linseed, and hay, ...	108	3.48	4.14
6. Beans and hay, . . .	108	3.72	5.27

Here grass produced the largest flow of milk, but of a quality less rich than bean-meal and hay, which produced the richest quality ; one hundred and eight pounds making more butter than one hundred and fourteen pounds of grass-made milk.

In autumn, the best feed will be the grasses of the pastures, so far as they are available, green-corn fodder, cabbage, carrot and turnip leaves, and an addition of meal or shorts. Towards the middle of autumn, the cows fed in the pastures will require to be housed regularly nights, especially in the more northern latitudes, and put, in part at least, upon hay. But every farmer knows that it is not judicious to feed out the best part of his hay when his cattle are first put into the barn, and that he should not feed so well in the early part of winter that he cannot feed better as it advances.

At the same time, it should always be borne in mind that the change from grass to a poor quality of hay or straw, for cows in milk, should not be too sudden. A poor quality of dry hay is far less palatable in the early part of winter, after the cows are taken from grass, than at a later period ; and, if it is resorted to with milch cows, will inevitably lead to a falling off in the milk, which no good feed can afterwards wholly restore.

It is desirable, therefore, to know what can be used instead of his best English or upland meadow hay, and yet not suffer any greater loss in the flow of milk, or condition, than is absolutely necessary. In some sections of New England, the best quality of swale hay will be used ; and the composition of that is as variable as possible, depending on the varieties of grasses of which it was made, and the manner of curing. But in other sections, many will find it necessary to use straw, and other substitutes; and it may be desirable to know how much is required to form an equivalent in nutrition to good meadow or English hay. The follow-ing brief table of nutritive equivalents will be conve-nient for reference:

	Nutritive equivalent.	Percentage of Nitrogen.	
		Dried.	Undried.
1 Meadow hay, ...	100	1.34	1.15
2 Red Clover-hay, ...	75	1.70	1.54
3. Rye-straw, ...	479	0.30	0.24
4. Oat-straw, ...	383	0.36	0.30
5. Wheat-straw, ...	426	0.36	0.27
6. Barley-straw, ...	460	0.30	0.25
7. Pea-straw, ...	64	1.45	1.79

The following is the composition of these several substances, in which their relative value will more distinctly appear :

Water.	Woody fibre.	Starch, Gum, Sugar.	Gluten, Albu-men, etc.	Fatty matter.	Saline matter.
14	30	40	7.1	2 to 5	5 to 10
14	25	40	9.3	3 to 5	9
12 to 15	45	38	1.3		4
12	45	35	1.3	0.8	6
12 to 15	50	30	1.3	2 to 3	5
12 to 15	50	30	1.3		5
10 to 15	25	45	12.3	1.5	4 to 6

From these tables it will be seen that, taking good English or meadow hay as the standard of comparison, and calling that one, 4.79 times the weight of rye-straw, or 3.83 times the weight of oat-straw, contains the same amount of nutritive matter; that is, it would take 4.79 times as much rye-straw to produce the same result as good meadow hay.

The more elaborate nutritive equivalents of Boussin-gault will be found to be very valuable and suggestive, and the following table is given in this connection for the sake of convenient reference.

11*

NUTRITIVE EQUIVALENTS. (PRACTICAL AND THEORETICAL.)

ARTICLES OF FOOD.	THEORETICAL VALUES.						Practical values, as obtained by experiments in feeding, according to						
	BOUSSINGAULT.				FRESENIUS.		Block.	Petri.	Meyer.	Thaer.	Pabst.	Schwartz.	Schweitzer.
	Water in 100 parts.	Nitrogen in 100 parts of dried substance.	Nitrogen in 100 parts of undried substance.	Nutritive equivalent.	Relative proportion of nitrogenized to non-nitrogenized substances.	Nutritive equivalent.							
English Hay,	11.0	1.34	1.15	100		100	100	100	100	100	100	100	100
Lucerne,	16.6	1.66	1.38	83			100	90		90	100	100	
Red Clover-hay,	10.1	1.70	1.54	75		77.9	430	90		90	100	100	
Red Clover (green),	76.0		.64	311	1 to 6.08		200		150	450	425		267
Rye-straw,	18.7	.30	.24	479		527 7-12	200	500	150	666	350	400	200
Oat-straw,	21.0	.36	.30	383	1 to 24.40	445 5-12		200		190	200		
Carrot-leaves (tops),	70.9	2.94	.85	135	1 to 12.50								
Swedish Turnips,	91.0	1.83	.17	676					250	300	250	200	
Mangold Wurzel,					1 to 7.26	391½	366	300		460	250	333	366⅔
White Silician Beet,	85.6	1.43	.18	669	1 to 7.84	542.1	366	400					
Carrots,	87.6	2.40	.30	382	1 to 9.00	330 5-12	216	250	225	300	250	270	300
Potatoes,	75.9	1.50	.36	319			400	200	150	200	200	200	200
Potatoes kept in pits,	76.8	1.18	.30	383									
Beans,	7.9	5.50	5.11	23	1 to 2.8	34 5-12	30	54	50	73	40		30
Peas,	8.6	4.20	3.84	27	1 to 2.14	34½	30	54	48	66	40	Boussingault 59	30
Indian Corn,	18.0	2.00	1.64	70	1 to 6.55			52					
Buckwheat,	12.5	2.40	2.10	55	1 to 6.05	93 5-12	33	64	53	76	50		35
Barley,	13.2	2.02	1.76	65	1 to 4.25		39½	61		86	60		37½
Oats,	12.4	2.22	1.92	60	1 to 4.08	58 11-12	33	71		71	50		33½
Rye,	11.5	2.27	2.00	58	1 to 4.42	58 1-16	27	55	51	64	40		30
Wheat,	10.5	2.33	2.09	55	1 to 2.42	38 5-6	42	52	46				43
Oil-cake (Linseed),	13.4	6.00	5.20	22				108					

The reader will find no difficulty in making this table of practical value in deciding upon the proper course of feeding to be pursued.

In winter the best food for cows in milk will be good sweet meadow hay, a part of which should be cut and moistened with water, as all inferior hay or straw should be, with an addition of root-crops, such as turnips, car-rots, parsnips, potatoes, mangold wurzel, with shorts, oil-cake, Indian-meal, or bean-meal.

It is the opinion of most successful dairymen that the feeding of moist food cannot be too highly recommended for cows in milk, especially to those who desire to obtain the largest quantity. Hay cut and thoroughly moistened becomes more succulent and nutritive, and partakes more of the nature of green grass.

As a substitute for the oil-cake, hitherto known as an exceedingly valuable article for feeding stock, there is probably nothing better than cotton-seed meal, now to be had in large quantities in the market. This is an article whose economic value has been but recently made known, but which, from practical trials already made, has proved eminently successful as food for milch cows. An average specimen of this was submitted for analysis to Professor Johnson, who reported that its composition is not inferior to that of the best flax-seed cake, and that in some respects its agricultural value surpasses that of any other kind of oil-cake, as is shown in the following table, containing in column first the analysis of cotton-seed meal made by himself; in column second, some of the results obtained by Dr. C. T. Jack-son on cake prepared by himself from hulled cotton-seed; in column third, an analysis of cotton-seed cake, made by Dr. Anderson, of Edinburgh; in column fourth, the aver-age composition of eight samples of American linseed-cake; and in column fifth, an analysis of meadow hay.

obtained by Dr. Wolff in Saxony, given as a means of comparison.

	I.	II.	III.	IV.	V.
Water, ...	6.82		11.19	9.23	16.94
Oil, ...	16.47	–	9.08	12.96	–
Albuminous bodies, . .	44.41	48.82	25.16	28.28	10.69
Mucilaginous and Saccharine matters, ...	} 12.74	{ . . .	} 48.93	34.22	40.11
Fibre, ...	11.76			9.00	27.16
Ash, ...	7.80	8.96	5.64	6.21	5.04
	100.00		100.00	100.00	100.00
Nitrogen, ...	7.05	7.75	3.95	4.47	–
Phosphoric acid in ash, .	2.36	2.45	–	–	–
Sand, ...	.94	–	1.32	–	–

Johnson also remarks, in this connection, that the great value of linseed-cake, as an adjunct to hay for fat cattle and milch cows, has long been recognized; and is undeniably traceable in the main to three ingredients of the seeds of the oil-yielding plants. The value of food depends upon the quantity of matters it contains which may be appropriated by the animal which consumes the food. Now, it is proved that the fat of animals is derivable from the *starch, gum,* and *sugar,* and more directly and easily from the *oil* of the food. These four substances are, then, the *fat-formers.* The muscles, nerves, and tendons of animals, the fibrine of their blood, and the curd of their milk, are almost identical in composition, and strongly similar in many of their properties with matters found in all vegetables, but chiefly in such as form the most concentrated food. These *blood* (and muscle) *formers* are characterized by containing about fifteen and a half per cent. of nitrogen; and hence are called *nitrogenous substances.* They are also often designated as the *albuminous bodies.*

The bony framework of the animal owes its solidity to *phosphate of lime,* and this substance must be fur nished by the food. A perfect food must supply the animal with these three classes of bodies, and in proper proportions. The addition of a small quantity of a food rich in oil and albuminous substances to the ordinary kinds of feed, which contain a large quantity of vegetable fibre or woody matter, more or less indigestible, but nevertheless indispensable to the herbivorous animals, their digestive organs being adapted to a bulky food, has been found highly advantageous in practice. Neither hay alone nor concentrated food alone gives the best results. A certain combination of the two presents the most advantages.

A Bavarian farmer has recently announced that heifers fed, for three months before calving, with a little linseed-cake, in addition to their other fodder, acquire a larger development of the milk-vessels, and yield more milk afterwards, than similar animals fed as usual. Cotton-seed cake must have an equally good effect.

Some of those who have used cotton-seed cake have found difficulty in inducing cattle to eat it. By giving it at first in small doses, mixed with other palatable food, they soon learn to eat it with relish.

On comparing the analyses II. and I. with the average composition of linseed-cake IV., it will be seen that the cotton-seed cake is much richer in oil and albuminous matters than the linseed-cake. A correspondingly less quantity will therefore be required. Three pounds of this cotton-seed cake are equivalent to four of linseed-cake of average quality.

During the winter season, as already remarked, a frequent change of food is especially necessary, both as contributing to the general health of animals, and as a means of stimulating the digestive organs, and thus increasing the secretion of milk. A mixture used as cut feed, and well moistened, is now especially beneficial, since concentrated food, which would otherwise be given in small quantities, may be united with larger quantities of coarser and less nutritive food, and the complete assimilation of the whole be better secured. On this subject Dr. Voelcker truly observes that the most nutritious kinds of food produce little or no effect when they are not digested by the stomach, or if the digested food is not absorbed by the lymphatic vessels, and not assimilated by the various parts of the body. Now, the normal functions of the digestive organs not only depend on the composition of the food, but also on its volume. The volume or bulk of the food contributes to the healthy activity of the digestive organs, by exercising a stimulating effect on the nerves which govern them. Thus the whole organization of ruminating animals necessitates the supply of bulky food, to keep the animal in good condition.

Feed sweet and nutritious food, therefore, regularly, frequently, and in small quantities, and change it often, and the best results may be confidently expected. If the cows are not in milk, but are to come in in the spring, the difference in feeding should be rather in the quantity than the quality, if the highest yield is to be expected from them the coming season.

The most common feeding is hay alone, and oftentimes very poor hay, at that. The main point is to keep the animal in a healthy and thriving condition, and not to suffer her to fail in flesh; and with this object some change and variety of food is highly important. And here it may be remarked that cows in calf should not, as a general rule, be milked the last month or six weeks before calving, and many prefer to have them run dry as many as eight or ten weeks. The yield of milk is better the coming season, and holds out better, than if they are milked up to the time of calving.

There are exceptions, however, and it is often very difficult to dry off a cow sufficiently to make it judicious to cease milking much, if any, before the time of calving. Some even prefer to milk quite up to this time; but the weight of authority among the best practical farmers is so decidedly against it, that there can be no question of its bad economy. Towards the close of winter, a herd of cows will begin to come in, or approach their time of calving. Care should then be taken not to feed too rich or stimulating food for the last week or two before this event, as it is often attended with ill consequences. A plenty of hay, a few potatoes or shorts, and pure water, will be sufficient.

As the time of calving approaches, the cow should be removed from the rest of the herd, to a pen with a level floor, by herself. Nothing is needed, usually, but to supply her regularly with food and drink, and leave her quietly to herself. In most cases the parturition will be natural and easy, and the less the cow is disturbed or meddled with, the better. She will do better without help than with; but she should be watched, in order to see that no difficulty occurs which may require aid and attention. In cases of difficult parturition the aid of a skilful veterinary surgeon may be required. For those who may desire to make themselves familiar with the details of such cases so far as to be able to act for themselves, Skellctt's "Practical Treatise on the Parturition of the Cow, or the Extraction of the Calf," an elaborate work, published in London in 1844, will be an important guide.

In spring the best feeding for dairy cows will be much the same as that for winter; the roots in store over winter, such as carrots, mangold wurzel, turnips, and parsnips, furnishing very valuable aid in increasing the quantity and improving the quality of milk. Tow-

ards the close of this season, and before the grass of
the pastures is sufficiently grown to make it judicious
to turn out the cows, the best dairymen provide a supply of green fodder in the shape of winter rye, which,
if cut while it is tender and succulent, and before it is
half grown, will be greatly relished. Unless cut young,
however, its stalk soon becomes hard and unpalatable.

Having stated briefly the general principles of feeding cows for the dairy, it is proper to give the statements of successful practical dairymen, both as corrob-
orating what has already been said, and as showing the
difference in practice in feeding and managing with
reference to the specific objects of dairy farming. And
first, a farmer of Massachusetts, supplying milk for the
Boston market, and feeding for that object, says: " For
thirty cows, cut with a machine thirty bushels for one
feed; one third common English hay, one third salt hay,
and one third rye or barley straw; add thirty quarts of
wheat bran or shorts, and ten quarts of oat and corn
meal moistened with water. One bushel of this mixture
is given to each cow in the morning, and the same
quantity at noon and in the evening. In addition to
this, a peck of mangold wurzel is given to each cow
per day. This mode of feeding has been found to produce nearly as much milk as the best grass feed in summer. When no wheat-bran or any kind of meal is given,
the hay is fed without cutting."

Another excellent farmer, of the western part of the
same state, devoting his attention to the manufacture
of cheese, and the successful competitor for the first
prize of the state society for dairies, says of his feeding:
" My pastures are upland, and yield sweet feed. I fed,
in the month of June, all the whey from the milk made
into cheese, without any meal. In September, my pastures being very much dried up, I fed all the whey.

with one quart of meal to each cow, and also ten pounds of corn fodder to each cow per day.

" I commence feeding my cows in the spring, before calving, with three quarts of meal each per day, until the feed in the pasture is good.

" I consider the best mixture of grain, ground into meal, for milk, is equal quantities of rye, buckwheat, and oats. For the last ten years I have not made less than five hundred pounds of cheese and twenty pounds of butter to each cow ; and one year I made six hundred and forty pounds of cheese and twenty pounds of butter to each cow.

" A cow will give more milk on good fresh grass than any other feed. When the grass begins to fail, I make up the deficiency by extra feed of meal and corn fodder. I feed all my whey to my cows. I let them run dry four months, and during this time I give them no extra feed, always keeping salt before them."

Another, with one of the best butter dairies in the same state, explains his mode of management of cows in the stall as follows : " In the management of my stock the utmost gentleness is observed, and exact regularity in the hours of feeding while confined to the stable, and of milking throughout the year.

" The stock is fed regularly three times a day.

" In the morning, as soon as the milking is over, each cow (having been previously fed, and her bag cleaned by washing, if necessary) is thoroughly cleaned and groomed, if the expression may be used, with a curry-comb, from head to foot, and, when cleaned, turned out to drink. The stable is now cleaned out, the mangers swept, and the floors sprinkled with plaster ; and as the cows return, which they do as soon as inclined, they are tied up and left undisturbed until the next hour of feeding, which is at noon.

12

"The cattle at this time are again turned out to drink, and, after being tied up on their return again fed. Of course the stable is at this time again ther oughly cleansed. And so again at night the same course is pursued. At this time a good bedding is spread for each cow, and, after all are in, they are fed.

"At six o'clock the milking commences, and at its termination, after removing from the floor whatever manure may have been dropped, the stable is closed for the night. If carrots are fed, which is the only root allowed to my cows in milk, they are given at the time of the evening milking.

"Whatever material is taken for bedding (as corn-stalks, husks, &c.) is passed through a cutting-machine. and composes the noon feed, such portions as are not consumed by the cows being used for bedding. The additional labor of cutting up is amply compensated by the reduced amount of labor in working (loading) and ploughing under the manure.

"While I consider it highly desirable that the cows, during the period they are stabled, should be kept warm and dry, I regard it as indispensable that they should be perfectly clean; and, although the stock is stabled the whole time, care is taken that there is a sufficient degree of ventilation."

In Herkimer county, New York, one of the best dairy districts in the country, a dairy farmer who kept twenty-five cows for the manufacture of cheese, making in one year nearly seven hundred pounds per cow, states his mode of feeding as follows : "When the ground is set-tled, and grass is grown so that cows can get their fill without too much toil, they are allowed to graze an hour, only, the first day ; the second day a little longer, and so on, till they get accustomed to the change of feed before they are allowed to have full range of pasture. Shift of pasture is frequently made to keep feed fresh and a good bite. About one acre per cow affords plenty of feed till the first of August. If enough land was turned to pasture to feed the cows through the season, it would get a start of them about this time, and be hard and dry the balance of the season. To avoid turning on my meadows in the fall, I take one acre to every ten cows, plough and prepare it the fore part of June for sowing; I commence sowing *corn* broadcast, about half an acre at a time (for twenty-five cows), so that it may grow eighty or ninety days before it is cut and fed. I have found, by experiment, that it then contains the most saccharine juice, and will produce the most milk. If the ground is strong, I sow two bushels per acre; more if the ground is not manured.

" The common yield is from fifteen to twenty tons (of green feed) per acre. About the first of August, when heat and flies are too oppressive for cows to feed quietly in the day-time, I commence feeding them with what corn they will eat in the morning, daily, which is cut up with a grass-scythe, and drawn on a sled or wagon to the milk-barn and fed to them in the stalls, which is one hour's work for a man at each feeding. When thus plentifully fed, my cows have their *knitting*-work on hand for the day, which they can do up by lying quietly under artificial shades, erected in such places as need manuring most, and are most airy, by setting posts and putting poles and bushes on top, the sides being left open. These shades may be made and removed annu-ally, to enrich other portions of soil, if desired, at the small expense of one dollar for each ten cows. At evening, my cows are fed whey only, because they can feed more quietly, with less rambling, and will give more milk by feeding most when the dew is on the grass.

"The capacity of cows for giving milk is varied much by habit. In fall, after the season of feeding is past, I feed four quarts of wheat bran or shorts made into slop with whey, or a peck of roots to each cow, till milking season closes (about the first of December). When confined in stables and fed hay and milked, they are fed each one pail full of thin slop at morning before foddering, and also at evening, to render their food more succulent, and they will not drink so much cold water when let out in the middle of the day. In cold weather cows are kept well attended in warm stables. No foddering is done on the ground. Thus a supply of milk is kept up, and the cows get in good flesh, while their blood and bags are left in a healthy condition when dried off.

" This flesh they hold till milk season in spring, without other feed than good hay. They will not get fleshy bags, but come into milk at once. About the first of April they are carded daily, till they are turned to grass. Wheat-bran in milk or whey, slops, or roots, are daily fed, as they are found best adapted to the nature of different cows, and most likely to establish a regular flow of milk till grass comes."

All practical dairymen concur in saying that a *warm and well-ventilated* barn is indispensable to the promotion of the highest yield of milk in winter ; and most agree that cows in milk should not be turned out even to drink in cold weather, all exposure to cold tending to lessen the yield of milk.

In the London dairies, where, of course, the cows are fed so as to produce the largest flow of milk, the treatment is as follows : The cows are kept at night in stalls. About three A. M. each has half a bushel of grains. When milking is finished, each receives a bushel of turnips (or mangolds), and shortly afterwards one tenth of a truss of hay of the best quality. This feeding occurs before eight A. M., when the animals are turned into the yard. Four hours after, they are again tied up in their stalls, and have another feed of grains. When the afternoon milking is over (about three P. M.), they are fed with a bushel of turnips, and after the lapse of an hour, hay is given them as before. This mode of feeding usually continues throughout the root season, or from November to March. During the remaining months they are fed with grains, tares, and cabbages, and a proportion of rowen or second-cut hay. They are supplied regularly until they are turned out to grass, when they pass the whole of the night in the field. The yield is about six hundred and fifty gallons a year for each cow.

Mr. Harley, whose admirable dairy establishment has been already alluded to, as erected for the purpose of supplying the city of Glasgow with a good quality of milk, and which contributed more than anything else to improve the quality of milk furnished to all the cities of Great Britain, adopted the following system of feeding with the greatest profit: In the early part of summer, young grass and green barley, the first cutting especially, mixed with a large proportion of old hay or straw, and a good quantity of salt to prevent swelling, were used. As summer advanced less hay and straw were given, and as the grass approached ripeness they were discontinued altogether, but young and wet clover was never given without an admixture of dry provender. When grass became scarce, young turnips and turnip-leaves were steamed with hay, and formed a good substitute. As grass decreased the turnips were increased, and at length became a complete substitute. As the season advanced a large proportion of distillers' grains and wash was given with

12*

other food, but these were found to be apt to make the
cattle grain-sick; and if this feeding were long continued, the health of the cows became affected. Boiled
linseed and short-cut wheat-straw mixed with the
grains were found to prevent the cows from turning
sick. As spring approached, Swedish turnips, when
cheap, were substituted for yellow turnips. These two
roots, steamed with hay and other mixtures, afforded
soft food till grass was again in season. When any of
the cows were surfeited, the food was withheld till the
appetite returned, when a small quantity was given,
and increased gradually to the full allowance.

But the most elaborate and valuable experiments in
the feeding and management of milch cows are those
recently made by Mr. T. Horsfall, of England, and published in the Journal of the Royal Agricultural Society.
His practice, though adapted, perhaps, more especially
to his own section, is nevertheless of such general
application and importance as to be worthy of attention.
By his course of treatment he found that he could produce as much and as rich butter in winter as in
summer.

His first object was to afford a full supply of the elements of food adapted to the *maintenance* and also to
the *produce* of the animal; and this could not be effected
by the ordinary food and methods of feeding, since it is
impossible to induce a cow to consume a quantity of
hay requisite to supply the waste of the system, and
keep up, at the same time, a full yield of the best
quality of milk. He used, to some extent, cabbages,
kohl rabi, mangolds, shorts, and other substances, rich
in the constituents of cheese and butter. " My food for
milch cows," says he, " after having undergone various
modifications, has for two seasons consisted of rape-cake
five pounds and bran two pounds, for each cow. mixed

with a sufficient quantity of bean-straw, oat-straw, and shells of oats, in equal proportions, to supply them three times a day with as much as they will eat. The whole of the materials are moistened and blended together, and, after being well steamed, are given to the animals in a warm state. The attendant is allowed one pound to one and a half pounds per cow, according to circumstances, of bean-meal, which he is charged to give to each cow in proportion to the yield of milk; those in full milk getting two pounds each per day, others but little. It is dry, and mixed with the steamed food on its being dealt out separately. When this is eaten up, green food is given, consisting of cabbages from October to December, kohl rabi till February, and mangold till grass time. With a view to nicety of flavor, I limit the supply of green food to thirty or thirty-five pounds per day for each. After each feed, four pounds of meadow hay, or twelve pounds per day, is given to each cow. They are allowed water twice a day to the extent they will drink."

Bean-straw uncooked being found to be hard and unpalatable, it was steamed to make it soft and pulpy, when it possessed an agreeable odor, and imparted its flavor to the whole mess. It was cut for this purpose just before ripening, but after the bean was fully grown, and in this state was found to possess nearly double the amount of albuminous matter, so valuable to milch cows, of good meadow or upland hay. Bean or shorts is also vastly improved by steaming or soaking with hot water, when its nutriment is more readily assimilated. It contains about fourteen per cent. of albumen, and is rich in phosphoric acid. Rape-cake was found to be exceedingly valuable. Linseed and cotton-seed cake may probably be substituted for it in this country. Mr. Horsfall is accustomed to turn his cows in May into a rich pasture, housing them at night, and giving them a mess of the steamed mixture and some hay morning and night; and from June to October they have cut grass in the stall, besides what they get in the pasture, and two feeds of the steamed mixture a day. After the beginning of October the cows are kept housed. With such management, his cows generally yield from twelve to sixteen quarts of milk (wine measure) a day, for about eight months after calving, when they fall off in milk, but gain in flesh, up to calving-time. In this course of treatment the manure is far better than the average, and his pastures are constantly improved. The average amount of butter from every sixteen quarts of milk is twenty-five ounces, a proportion far larger than the average. His investigations are very full and complete. — See Appendix.

How widely does this course of practice differ from that of most farmers ! The object with many seems to be to see with how little food they can keep the cow alive. Now, it appears to me that the milch cow should be regarded as an instrument of transformation. With so much hay, so much grain, so many roots, how can the most milk, or butter, or cheese, be made? The conduct of a manufacturer who owned good machinery, and an abundance of raw material, and had the labor at hand, would be considered as very absurd, if he hesitated to supply the material, and keep the machinery at work at least so long as he could run it with profit.

Stimulate the appetite, then, and induce the cow to eat, by a frequent change of diet, not merely enough to supply the constant waste of her system, but enough and to spare, of a food adapted to the production of milk of the quality desired.

SOILING. — Of the advantages of soiling milch cows, or feeding exclusively in the barn, there are still many conflicting opinions. As to its economy of land and feed there is no question, it being generally admitted that a given number of animals may be abundantly fed on a less space; nor is there much question as to the increased quantity of milk yielded in stall feeding. Its economy in this country turns rather upon the cost of labor and land; and the question asked by the dairy-man is whether it will pay — whether its advantages are sufficient to balance the extra expense of cutting and feeding over and above cropping on the pasture. The importance of this subject has been strongly im-pressed upon the attention of farmers in many sections of the country, by a growing conviction that something must be done to improve the pastures, or that they must be abandoned altogether.

Thousands of acres of neglected pasture-land in the older states are so poor and worn out that from four to eight acres furnish but a miserable subsistence for a good-sized cow. No animal can flourish under such cir-cumstances. The labor and exertion of feeding is too great, to say nothing of the vastly inferior quality of the grasses in such pastures to those on more recently seeded lands. True economy would dictate that such pastures should either be allowed to run up to wood, or be devoted to sheep-walks, or ploughed and improved. Cows, to be able to yield well, must have plenty of food of a sweet and nutritious quality; and unless they find it, they wander over a large space, if at liberty, and deprive themselves of rest.

If a farmer or dairyman is the unfortunate owner of such pastures, there can be no question that, as a mat-ter of real economy, he had better resort to the soiling system for his milch cows, by which means he will largely increase his annual supply of good manure, and thus have the means of improving, and bringing his

land to a higher state of cultivation. A very successful instance of this management occurs in the report of
the visiting committee of an agricultural society in
Massachusetts, in which they say : " We have now in
mind a farmer in this county who keeps seven or eight
cows in the stable through the summer, and feeds them
on green fodder, chiefly Indian-corn. We asked him
the reasons for it. His answer was : 1. That he gets
more milk than he can by any other method. 2. That
he gets more manure, especially liquid manure. 3.
That he saves it all, by keeping a supply of mould or
mud under the stable, to be taken out and renewed as
often as necessary. 4. That it is less troublesome than
to drive his cows to pasture ; that they are less vexed
by flies, and have equally good health. 5. That his
mowing-land is every year growing more productive,
without the expense of artificial manure. He estimates
that on an acre of good land twenty tons of green fodder may be raised. That which is dried is cut fine, and
mixed with meal or shorts, and fed with profit. He
believes that a reduced and partially worn-out farm —
supposing the land to be naturally good — could be
brought into prime order in five years, without extra
outlay of money for manure, by the use of green fodder in connection with the raising and keeping of pigs;
not fattening them, but selling at the age of four or five
months." He keeps most of his land in grass, improving its quality and productiveness by means of top-
dressing, and putting money in his pocket, — which is,
after all, the true test both for theory and practice.

Another practical case in hand on this point is that
of a gentleman in the same state, who had four cows,
but not a rod of land to pasture them on. They were,
therefore, never out of the barn, — or, at least, not
out of the yard, — and were fed with grass, regularly

mown for them ; with green Indian-corn fodder, which
had been sown broadcast for the purpose ; and with
about three pints of meal a day. Their produce in butter was kept for thirteen weeks. Two of them were
but two years old, having calved the same spring. All
the milk of one of them was taken by her calf six
weeks out of the thirteen, and some of the milk of the
other was taken for family use, the quantity of which
was not measured. These heifers could not be estimated, therefore, as more than equal to one cow in full
milk. And yet from these cows no less than three
hundred and eighty-nine pounds of butter were made in
the thirteen weeks. Another pound would have made
an average of thirty pounds a week for the whole time.

It appears from these, and other similar instances of
successful soiling, or stall-feeding in summer on green
crops cut for the purpose, that the largely increased
quantity of the yield fully counterbalances the slightly
deteriorated quality. And not only is the quantity
yielded by each cow increased, but the same extent of
land, under good culture, will carry double or treble the
number of ordinary pastures, and keep them in better
condition. There is also a saving of manure. But with
us the economy of soiling is the exception, and not the
rule.

In adopting this system of feeding, regularity is
required as much as in any other, and a proper variety
of food. A succession of green crops should be
provided, as near as convenient to the stable. The
first will naturally be winter rye, in the Northern
States, as that shoots up with great luxuriance. Winter rape would probably be an exceedingly valuable
addition to the plants usually cultivated for soiling
in this country, in. sections where it withstands the
severity of the winter. Cabbages kept in the cellar, or

pit, and transplanted early, will also come in here to advantage, and clover will very soon follow them; oats, millet, and green Indian-corn, as the season advances; and, a little later still, perhaps, the Chinese sugar-cane, which should not be cut till headed out. These plants, in addition to other cultivated grasses, will furnish an unfailing succession of succulent and tender fodder; while the addition of a little Indian, linseed, or cotton-seed meal will be found economical.

In the vicinity of large towns and cities, where the object is too often to feed for the largest quantity, without reference to quality, an article known as distillers' swill, or still-slop, is extensively used. This, if properly fed in limited quantities, in combination with other and more bulky food, may be a valuable article for the dairyman ; but, if given, as it too often is, without the addition of other kinds of food, it soon affects the health and constitution of the animals fed on it. This swill contains a considerable quantity of water, some nitrogenous compounds, and some inorganic matter, in the shape of phosphates and alkaline salts found in the different kinds of grain of which it is made up, as Indian corn, wheat, barley, rye, &c. Where this forms the principal food of milch cows, the milk is of a very poor quality — blue in color, and requiring the addition of coloring substances to make it salable. It contains, often, less than one per cent. of butter, and seldom over one and three tenths or one and a half per cent., while good, salable milk ought to contain from three to five per cent. It will not coagulate, it is said, in less than five or six hours, while good milk will invariably coagulate in one hour or less, under the same conditions. Its effect on the system of young children is therefore very destructive, causing diseases of various kinds, and, if continued, certain death.

MILKING.—The manner of milking exerts a more powerful and lasting influence on the productiveness of the cow than most farmers are aware of. That a slow and careless milker soon dries up the best of cows, every practical farmer and dairyman knows ; but a care· ful examination of the beautiful structure of the udder will serve further to explain the proper mode of milking, to obtain and keep up the largest yield. " The udder of a cow," says a writer in the Rural Cyclopædia, " is a unique mass, composed of *two* symmetrical parts, simply united to each other by a cellular tissue, lax, and very abundant; and each of these parts comprises two divisions or quarters, which consist of many small granules, and are connected together by a compact laminous tissue; and from each quarter proceed systems of ducts, which form successive unions and confluences, somewhat in the manner of the many affluents of a large river, until they terminate in one grand excretory canal, which passes down through the elongated mam- millary body called the teat. Its lactiferous or milk tubes, however, do not, as might be supposed, proceed exactly from smaller to larger ducts by a gradual and regular enlargement, because it would not have been proper that the secretion of milk should escape as it was formed; and therefore we find an apparatus ádapted for the purpose of retaining it for a proper time. This apparatus is to be found both in the teat and in the in- ternal construction of the udder. The teat resembles a funnel in shape, and somewhat in office; and it is pos- sessed of a considerable degree of elasticity. It seems formed principally of the cutis, with some muscular fibres, and it is covered on the outside by cuticle, like every other part of the body ; but the cuticle here not only covers the exterior, but also turns upwards, and lines the inside of the extremity of the teat, as far as it

13   10

is contracted, and there terminates by a frilled edge, the rest of the interior of the teats and ducts being lined by mucous membrane. But, as the udder in most animals is attached in a pendulous manner to the body, and as the weight of the column of fluid would press with a force which would, in every case, overcome the resistance of the contractions of the extremity, or prove oppressive to the teat, there is in the internal arrangement of the udder a provision made to obviate this difficulty. The various ducts, as they are united, do not become gradually enlarged so as to admit the ready flow of milk in a continual stream to the teat, but are so arranged as to take off, in a great measure, the extreme pressure to which the teat would be other-wise exposed. Each main duct, as it enters into another, has a contraction produced, by which a kind of valvular apparatus is formed in such a manner as to become pouches or sacks, capable of containing the great body of the milk. In consequence of this arrangement, it is necessary that a kind of movement upwards, or lift, should be given to the udder before the teat is drawn, to force out the milk; and by this lift the milk is dis-placed from these pouches, and escapes into the teat, and is then easily squeezed out; while the contractions, or pouches, at the same time resist, in a certain degree, the return or reflux of the displaced milk."

The first requisite of a good milker is, of course, the utmost cleanliness. Without this, the milk is unen-durable. The udder should, therefore, be carefully cleaned before the milking commences. The milker may begin gradually and gently, but should steadily increase the rapidity of the operation till the udder is emptied, using a pail sufficiently large to hold all, with out the necessity of changing. Cows are very sensi-tive, and the pail cannot be changed, nor can the milker stop or rise during the process of milking, without leading the cow more or less to withhold her milk. The utmost care should be taken to strip to the last drop, and to do it rapidly, and not in a slow and negligent manner, which is sure to have its effect on the yield of the cow. If any milk is left, it is reäbsorbed into the system, or else becomes caked, and diminishes the tendency to secrete a full quantity afterwards. Milking as dry as possible is especially necessary with young cows with their first calf, as the mode of milking, and the length of time to which they can be made to hold out, will have very much to do with their milking qualities as long as they live.

At the age of two or three years the milky glands have not become fully developed, and their largest development will depend very greatly upon the management after the first calf. Cows should have, therefore, the most milk-producing food ; be treated with constant gentleness ; never struck, or spoken harshly to, but coaxed and caressed ; and in ninety-nine cases out of a hundred they will grow up gentle and quiet. But harshness is worse than useless. Nothing does so much to dry a cow up, especially a young cow.

The longer the young cow, with her first and second calf, can be made to hold out, the more surely will this habit be fixed upon her. Stop milking her four months before the next calf, and it will be difficult to make her hold out to within four or six weeks of the time of calving afterwards. Induce her, if possible, by moist and succulent food, and by careful milking, to hold out even up to the time of calving, if you desire to milk her so long, and this habit will be likely to be fixed upon her for life. But do not expect to obtain the full yield of a cow the first year after calving. Some of the very best cows are slow to develop their best qualities ; and no cow reaches her prime till the age of five or six years.

The extreme importance of care and attention to these points cannot be over-estimated. The wild cows grazing on the plains of South America are said to give only about three or four quarts a day at the height of the flow; and many an owner of large herds in Texas, it is said, has too little milk for family use, and sometimes receives his supply of butter from the New York market. There is, therefore, a constant tendency to dry up in milch cows ; and it must be guarded against with special care, till the habit of yielding a large quantity, and yielding it long, becomes fixed in the young animal, when, with proper care, it may easily be kept up.

If gentle and mild treatment is observed and perse-vered in, the operation of milking appears to be one of pleasure to the animal, as it undoubtedly is; but if an opposite course is pursued, — if, at every restless move-ment, caused, perhaps, by pressing a sore teat, the animal is harshly spoken to, — she will be likely to learn to kick as a habit, and it will be difficult to overcome it ever afterwards. To induce quiet and readiness to give down the milk freely, it is better that the cow should be fed at milking-time with cut feed, or roots, placed within her easy reach.

I have never practised milking more than twice a day, because in spring and summer other farm-work was too pressing to allow of it; but there is no doubt that, for some weeks after calving, and in the height of the flow, the cows ought, if possible, to be milked regularly three times a day — at early morning, noon, and night. Every practical dairyman knows that cows thus milked give a larger quantity of milk than if milked only twice, though it may not be quite so rich; and in young cows, no doubt, it has a tendency to promote the development of the udder and milk-veins. A frequent milking stimulates an increased secretion, therefore, and ought never to be neglected in the milk-dairy, either in the case of young cows or very large milkers, at the height of the flow, which will ordinarily be for two or three months after calving.

The charge of this branch of the dairy should generally be intrusted to women. They are more gentle and winning than men. The same person should milk the same cow regularly, and not change from one to another, unless there are special reasons for it.

There being a wide difference in the quality as well as in the quantity of milk of different cows, no dairyman should neglect to test the milk of each new addition to his dairy stock, whether it be an animal of his own raising or one brought from abroad. A lactometer is a very convenient instrument here ; but any one can set the milk of each cow separately at first, and give it a fair and full trial, when the difference will be found to be great. Economy will dictate that the cows least adapted to the purpose should be disposed of, and their place supplied by better ones.

THE BARN. — The management of dairy stock requires a warm and well-ventilated barn or cow-room, in latitudes where it becomes necessary to stall-feed during several months of the year. This should be arranged in a manner suitable to keeping hay and other fodder dry and sweet, and with reference to the comfort and health of animals, and the economy of labor and manure. The size and finish will, of course, depend on the wants and means of the farmer or dairyman; but many little conveniences can be added at trifling cost.

The cow-room, Fig. 56, is given as an illustration merely of a convenient arrangement for a medium-sized dairy, and not as adapted to all circumstances or situ-

13*

ations. The barn stands, we will suppose, upon a side hill, or an inclined surface, where it is easy to have a cellar, if it is desired; and the cow-room, as shown in the figure, is in the second story, or directly over the cellar, the bottom of which should be somewhat dished, or lower in the middle than around the outer sides, and carefully paved or laid in cement.

The cow-room, as shown in the figure, is drawn on a scale of twenty feet to the inch. On the outside is represented an open shed, *m.* for carts and wagons to remain under cover, thirty feet by fifteen, while *l l l l l l* are bins for vegetables, to be filled through scuttles from the floor of the story above, and surrounded by solid walls. The area of this whole floor equals one hundred feet by fifty-seven. *k*, open space, and nearly on a level with the cow-chamber, through the door *p*. *s*, stairs to third story and to the cellar. *d d d*, passage next to the walls, five feet wide, and nine inches above the dung-pit. *e e e*, dung-pit, two feet wide, and seven inches below the floor where the cattle stand. The manure drops from this pit into the cellar below, five feet from the walls, and quite round the cellar. *c c c*, plank floor for cows, four feet six inches long. *b b b*, stalls for three yoke of oxen, on a platform five feet six inches long. *n n*, calf-pens, which may be used also for cows in calving. . *r r*, feeding-troughs for calves. The feeding-boxes are made in the form of trays, with partitions between them. Water comes in by a pipe, to cistern *a*. This cistern is regulated by a cock and ball, and the water flows by dotted lines, *o o o*, to the boxes, and each box is connected by lead pipes well secured from frost, so that, if desired, each animal can be watered without leaving the stall, or water can be kept constantly before it. A scuttle by which sweep- ings, etc., may be put through into the cellar, is seen

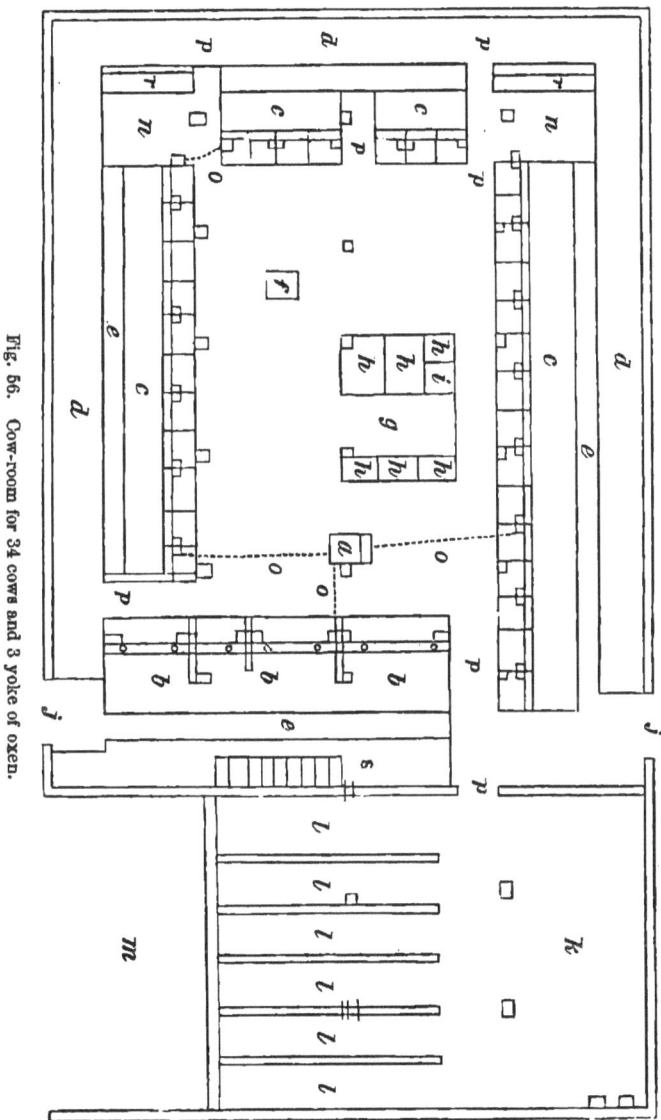

Fig. 56. Cow-room for 34 cows and 3 yoke of oxen.

at f. g is a bin receiving cut hay from third story, or hay-room. $h\,h\,h\,h\,h\,h$, bins for grain-feed. i is a tunnel to conduct manure or muck from the hay-floor to the cellar. $j\,j$, sliding doors on wheels. The cows all face towards the open area in the centre.

This cow-room may be furnished with a thermometer, clock, etc., and should always be well ventilated by sliding windows, which at the same time admit the light.

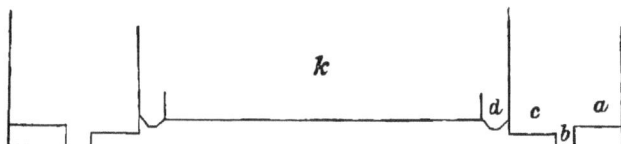

Fig. 57.

Fig. 57 is a transverse section of the cow-room, Fig. 56, *a* being a walk behind the cows, five feet wide; *b*, dung-pit; *c*, cattle-stand; *d*, feeding-trough, with a bottom on a level with the platform where the cattle stand; *k*, open area, forty-three feet by fifty-six.

The story above the cow-room, Fig. 58, is one hun dred feet by forty-two, the bays for hay, ten on each side, being ten feet front and fifteen feet deep, and the open space, *p*, for the entrance of wagons, carts, etc., twelve feet wide. *b*, hay-scales. *c*, scale-beam. *m m m m m m*, ladders reaching almost to the roof. *l l l*, &c., scuttle-holes for sending vegetables direct to the bins, *l l l*, etc., below. *a a b b*, rooms on the corners for storage. *d*, scuttles, four of which are used for straw, one for cut hay, and one for muck for the cellar. *n* and the other small squares are eighteen-feet posts. *f*, passage to the tool-house, a room one hundred feet long by fifteen wide. *o*, stairs leading to the scaffold in the roof of the tool-house. *i i*, benches. *g*, floor. *h*, boxes for hoes, shovels, spades, picks, iron bars, old iron, etc. *j j j*, bins for fruit. *k*, scuttles to put apples into wagons, etc., in the shed below. One side of this tool-house may be used for ploughs and large implements, hay-rigging, harrows, etc.

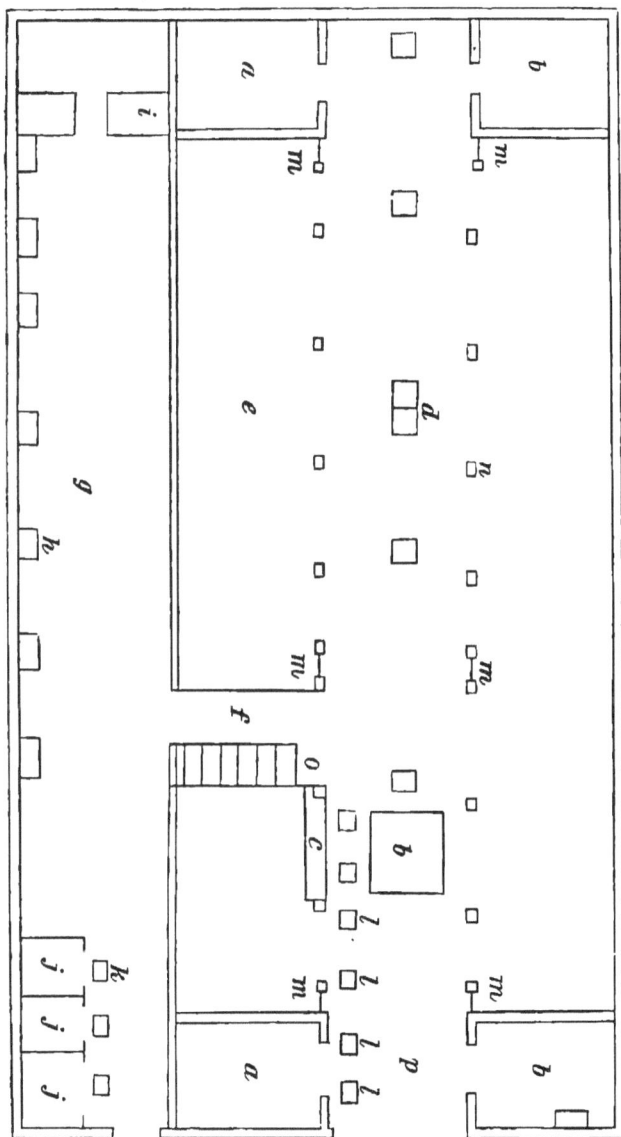

Fig. 58. Room over the cow-room for hay, &c.

Proper ventilation of the cellar and the cow-room
avoids the objection-that the hay is liable to injury
from noxious gases.

The excellent manure-cellar beneath this barn extends only under the cow-room. It has a drive-way through doors on each side. No barn-cellar should be kept shut up tight, even in cold weather. The gases are constantly escaping from the manure, unless held by absorbents, and are liable not only to affect the health of the stock, but to injure the quality of the hay. To prevent this, and yet secure the important advantages of a manure-cellar, the barn may be furnished with good-sized ventilators on the top, for every twenty-five feet of its length, and with wooden tubes leading from the cellar to the top.

There should also be windows on different sides of the cellar, to admit a free circulation of air. With these precautions, together with the use of absorbents in the shape of loam and muck, there will be no danger of rotting the timbers of the barn, or of risking the health of the cattle or the quality of the hay.

The temperature at which the cow-room should be kept is somewhere from 50° to 60°, Fahr. The practice and the opinions of successful dairymen differ on this point. Too great heat would affect the health and appetite of the herd, while too low a temperature is equally objectionable, for various reasons.

CHAPTER V.

It has been found in practice that calves properly bred and raised on the farm have a far greater intrinsic value for that farm, other things being equal, than any that can be procured elsewhere, while on the manner in which they are raised will depend much of their future usefulness and profit. These considerations should have their proper weight in the decision as to whether a promising calf from a good cow and bull shall be kept or sold to the butcher. But, rather than raise a calf at hap-hazard, and simply because its dam was celebrated as a milker, the judicious farmer will judge of the peculiar characteristics of the animal itself. This will often save a great and useless outlay which has sometimes been incurred in raising calves for dairy purposes, that a more careful examination would have rejected as unpromising.

The method of judging stock developed in a former chapter is of practical use here, and it is safer to rely upon it, to some extent, particularly when other appear-ances concur, than to go on blindly. The milk-mirror on the calf is small, but no smaller in proportion to its size than that of the cow; while its shape and form can generally be distinctly seen, particularly at the end of ten or twelve weeks. The development of the udder, and other peculiarities, will give some indication of the future capacities of the animal, and these should be studied.

If we except the manure of young stock, the calf is the first product of the cow, and as such demands our attention, whether it is to be raised or hurried off to the shambles. The practice adopted in raising calves differs widely in different sections of the country, being governed very much by local circumstances, as the vicinity of a milk-market, the value of milk for the dairy, the object of breeding, whether mainly for beef, for work, or for the dairy, etc. ; but, in general, it may be said that, within the range of thirty or forty miles of good veal-markets, which large towns furnish, comparatively few are raised at all. Most of them are fatted and sold at ages varying from three to eight or ten weeks; and in milk-dairies still nearer large towns and cities they are often hurried off at one or two days, or, at most, a week old. In both of these cases, as long as the calf is kept it is generally allowed to suckle the cow, and, as the treatment is very simple, there is nothing which particularly calls for remark, unless it be to condemn the practice entirely, on the ground that there is a more profitable way even for fattening calves for the butcher, and to say that allowing the calf to suck the cow at all is objectionable on the score of economy, except in cases where it is rendered necessary by the hard and swollen condition of the udder.

If the calf is so soon to be taken away, I should prefer not to suffer the cow to become attached to it at all, since she is apt to withhold her milk when it is removed, and a loss is sustained. The farmer will be governed by the question of profit, whatever course it is proposed to adopt. In raising blood stock, however, or in raising beef cattle, without any regard to economy of milk, the system of suckling the calves, or letting them run with the cow, may and will be adopted, since it is usually attended with somewhat less labor.

The other course, which is regarded as the best where the calf is to be raised for the dairy, is to bring it up by hand. This is done almost universally in all countries where the raising of dairy cows is best understood, —in Switzerland, Holland, some parts of Germany, and England. It requires rather more care, on the whole; but it is decidedly preferable, since the calves cost less, as the food can be easily modified, and the growth is not checked, as it is apt to be when the calf is finally taken off from the cow. I speak, of course, of sections where the milk of the cow is of some account for the dairy, and where it is too valuable to be devoted entirely to nourishing the calf. In this case, as soon as the calf is dropped the cow is allowed to lick off the slimy moisture till it is dry, which she will usually do from instinct, or, if not, a slight sprinkling of salt over the body of the calf will immediately tempt her. The calf is left to suck once or twice, which it will do as soon as it is able to stand. It should, in all cases, be permitted to have the first milk that comes from the cow, which is of a turbid, yellowish color, unfit for any of the purposes of the dairy, but somewhat purgative or medicinal, and admirably and wisely designed by nature to free the bowels and intestines of the new-born animal from the mucous, excrementitious matter always existing in them after birth. Too much of this new milk may, however, be hurtful even to the new-born calf, while it should never be given at all to older calves. The best course, it seems to me,—and I speak from considerable experience, and much observation and inquiry of others,—is to milk the cow dry immediately after the calf has sucked once, especially if the udder is painfully distended, which is often the case, and to leave the calf with the

cow during one day, and after that to feed it by putting
the fingers into its mouth, and gently bringing its
muzzle down to the milk in a pail or trough, when it
will imbibe in sucking the fingers. I have never found
much difficulty in teaching the calf to drink when taken
so young, though some take to it much more readily
than others. What the calf does not need should be
given to the cow. Some, however, prefer to milk
immediately after calving; and if the udder is over-
loaded this may be the best course, though the better
practice seems to be to leave the cow as quietly to herself as possible for a few hours. The less she is disturbed, as a general thing, the better. The after-birth
should be taken from her immediately after it is
dropped. It is customary to give the cow, as soon as
convenient, after calving, some warm and stimulating
drink, — a little meal stirred into warm water, with a
part of the first milk that comes from her, seasoned
with a little salt.

In many cases the calf is taken from the cow immediately, and before she has seen it, to a warm, dry pen
out of her sight, and there rubbed till thoroughly dry;
and then, when able to stand, fed with the new milk
from the cow, which it should have three or four times
a day, regularly, for the first fortnight, whatever course
it is proposed to adopt afterwards. It is of the greatest importance to give the young calf a thrifty start.
The milk, unless coming directly from the cow, should
be warmed.

Some object to removing the calf from the cow in this
way, on the ground of its apparent cruelty. But the
objection to letting the calf suckle the cow for several
days, as they do, or indeed of leaving it with the cow
for any length of time, is, that she invariably becomes
attached to it, and frets and withholds her milk when

it is at last taken from her. She probably suffers a great deal more, after this attachment is once formed, at the removal of the object of it, than she does at its being taken at once out of her sight. The cow's memory is far greater than many suppose; and the loss and injury sustained by removing the calf after it has been allowed to suck her for a longer or shorter period is never known exactly, because it is not usually known how much milk the calf takes; but it is, without doubt, very considerable. If the udder is all right, there seems to be no good reason for leaving the calf with the cow two or three days, if it is then to be taken away.

The practice in Holland is to remove the calf from the mother even before it has been licked, and to take it into one corner of the barn, or into another building, out of the cow's sight and hearing, put it upon soft dry straw, and rub it dry with some hay or straw, when its tongue and gums are slightly rubbed with salt, and the mucus and saliva removed from the nostrils and lips. After this has been done, the calf is made to drink the milk first taken as it comes from the mother. It is slightly diluted with water, if taken last from the udder; but, if the first of the milking, it is given just as it is. The calf is taught to drink in the same manner as in this country, by putting the fingers in its mouth and bringing it down to the milk, and it soon gets so as to drink alone. It is fed at first from four to six times a day, or even oftener; but soon only three times, at regular intervals. Its food for two or three weeks is clear milk, as it comes warm and fresh from the cow. This is never omitted, as the milk during the most of that time possesses certain qualities which are necessary to the calf, and which cannot be effectually supplied by any other food. In the third or fourth week the milk is skimmed, but warmed to the degree of fresh milk ; though, as the calf grows a little older, the milk is given cold, while less care is taken to give it the milk of its own mother, that of other cows now answering equally well.   In some places calves are fed on butter-milk at the age of two weeks and after ; but the change from new milk, fresh from the cow, is made gradually, some sweet skim-milk and warm water being at first added to it.

At three weeks old, or thereabouts, the calf will begin to eat a little sweet, fine hay, and potatoes cut fine, and it very soon becomes accustomed to this food. Many now begin to give linseed-meal mixed into hot water, to which is added some skim-milk or butter-milk; and others use a little bran cooked in hay-tea, made by chopping the hay fine, and pouring on boiling hot water, which is allowed to stand a while on it.   An egg is fre-quently broken into such a mixture.   Others still at this age take pains to have fresh linseed-cake, broken into pieces of the size of a pigeon's-egg; putting one of these into the mouth after the meal of milk has·been finished, and when it is eager to suck at anything in its way.   It will very soon learn to eat linseed-meal. A little sweet clover is put in its way at about the age of three weeks, and it will soon eat that also.

In this manner the feeding is continued from the fourth to the seventh week, the quantity of solid food being gradually increased.   In the sixth or seventh week the milk is by degrees withheld, and water or butter-milk used instead ; and soon after this, green food may be safely given, increasing it gradually with the hay to the age of ten or twelve weeks, when it will do to put them upon grass alone, if the season is favor-able for it.   A lot as near the house as possible, where they can be easily looked after and frequently visited, is best.   Calves should be gradually accustomed to all changes; and even after being turned to pasture they ought to be taken in if the weather is not dry and warm. The want of care and attention to these little details will be apparent sooner or later; while, if the farmer give his own time to these matters, he will be fully paid in the rapid growth of his calves. It is espe-cially necessary to see that the troughs from which they are fed, if troughs are used, are kept clean and sweet.

But there are some even among intelligent farmers who make a practice of turning their calves out to pasture at the tender age of two and three weeks, and that, too, when they have sucked the cow up to that time, and allow them nothing in the shape of milk or tender care. I cannot but think that this is the poorest possible economy, to say nothing of the cruelty of such treatment. The growth of the calf is checked, and the system receives a shock from so sudden a change, from which it cannot soon recover. The care-ful Dutch breeders bring the calves either skimmed milk or butter-milk to drink several times a day after they are turned to grass, which is not till the age of ten or twelve weeks; and, if the weather is chilly, the milk is warmed for them. They put a trough generally under a covering, where the calves may come and drink at regular times. Thus they are kept tame and docile.

In the raising of calves, through all stages of their growth, great care should be taken neither to starve nor to over-feed. A calf should never be surfeited, and never be fed so highly that it cannot be fed more highly as it advances. The most important point is to keep it growing thriftily without getting too fat, if it is to be raised for the dairy.

Mr. Aiton, in describing the mode of rearing calves in the dairy districts of Scotland, says: "They are fed on

milk, with seldom any admixture; and they are not permitted to suckle their dams, but are taught to drink
milk by the hand from a dish. They are generally fed
on milk only for the first four, five, or six weeks, and are
then allowed from two to two and a half quarts of new
milk each meal, twice in the twenty-four hours. Some
never give them any other food when young except
milk, lessening the quantity when the calf begins to
eat grass or other food, which it generally does
when about five weeks old, if grass can be had; and
withdrawing it entirely about the seventh or eighth
week of the calf's age. But, if the calf is reared
in winter, or early in spring, before the grass rises, it
must be supplied with at least some milk till it is eight
or nine weeks old; as a calf will not so soon learn to
eat hay or straw, nor fare so well on them alone as it
will do on pasture. Some feed their calves reared for
stock partly with meal mixed in the milk after the
third or fourth week. Others introduce gradually some
new whey among the milk, first mixed with meal;
and, when the calf gets older, they withdraw the milk,
and feed it on whey and porridge. Hay-tea, juices of
peas and beans, or pea or bean straw, linseed beaten
into powder, treacle, &c., have all been sometimes used
to advantage in feeding calves; but milk, when it can
be spared, is by far their most natural food.

"In Galloway, and other pastoral districts, where the
calves are allowed to suckle, the people are so much
wedded to their own customs as to argue that suckling is
much more nutritive to the calves than any other mode
of feeding. That suckling induces a greater secretion
of saliva, which, by promoting digestion, accelerates the
growth and fattening of the young animal, cannot be
doubted; but the secretion of that fluid may likewise
be promoted by placing an artificial teat in the mouth

of the calf, and giving it the milk slowly, and at the natural temperature. In the dairy districts of Scotland, the dairy-maid puts one of her fingers into the mouth of the calf, when it is fed, which serves the purpose of a teat, and will have nearly the same effect as the natural teat, in inducing the secretion of saliva. If that, or an artificial teat of leather, be used, and the milk given slowly before it is cold, the secretion of saliva may be promoted to all the extent that can be necessary; besides, that secretion is not confined to the mere period of eating, but, as in the human body, the saliva is formed and part of it swallowed at all times. As part of the saliva is sometimes seen dropping from the mouths of the calves, it might be advisable to give them not only an artificial teat, when fed, but to place, as is frequently done, a lump of chalk before them to lick, thus leading them to swallow the saliva. The chalk would so far supply the want of salt, of which cattle are so improperly deprived, and it would also promote the formation of saliva. Indeed, calves are much disposed to lick and suckle everything that comes within their reach, which seems to be the way that nature teaches them to supply their stomachs with saliva.

" But, though suckling their dams may be most advantageous in that respect, yet it has also some disadvantages. The cow is always more injured than the calf is benefited, by that mode of feeding. She becomes so fond of the calf that she does not, for a long time after yield her milk freely to the dairy-maid. The calf does not when young draw off the milk completely, and when it is taken off by the hand the cow withholds part of her milk; and, whenever a cow's udder is not completely emptied every time she is milked, the lactic secretion is thereby diminished.

" Feeding of calves by the hand is in various other

respects advantageous. Instead of depending on the
uncertain or perhaps precarious supply of the dam,
which may be more at first than the young animal can
consume or digest, and at other times too little for its
supply, its food can, by hand-feeding, be regulated to
suit the age, appetite, and purposes for which the calf
is intended ; other admixtures or substitutes can be
introduced into the milk, and the quantity gradually
increased or withdrawn at pleasure. This is highly
necessary when the calves are reared for stock. The
milk is in that case diminished, and other food introduced so gradually that the stomach of the young ani
mal is not injured as it is when the food is too suddenly
changed. And, in the case of feeding of calves for the
butcher, the quantity of milk is not limited to that of
the dam (for no cow will allow a stranger calf to suckle
her), but it can be increased, or the richest or poorest
parts of the milk given, at pleasure."

In these districts, where, probably, the feeding and
management of calves is as well and judiciously conducted as in any other part of Britain, the farmers'
wives and daughters, or female domestics, have the
principal charge of young calves ; and they are, no
doubt, much better calculated for this duty than men,
since they are more inclined to be gentle and patient.
The utmost gentleness should always be observed in
the treatment of all stock ; but especially of milch cows.
and calves designed for the dairy. Persevering kindness and patience will, almost invariably, overcome the
most obstinate natures ; while rough and ungentle handling will be repaid in a quiet kind of way, perhaps, by
withholding the milk, which will always have a tendency
to dry a cow up ; or, what is nearly as bad, by kicking,
and other modes of revenge, which often contribute to
the personal discomfort of the milker. The disposition

of the cow is greatly modified, if not, indeed, wholly
formed, by her treatment while young; and therefore
it is best to handle calves as much as possible, and
make pets of them, lead them with a halter, and caress
them in various ways. Calves managed in this way
will always be docile, and suffer themselves to be
approached and handled both in the pasture and the
barn.

With respect to the use of hay-tea, often used in this
country, but more common abroad, where greater care
and attention is usually given to the details of breeding,
Youatt says: " At the end of three or four days, *or perhaps a week, or even a fortnight,* after a calf has been
dropped, and the first passages have been cleansed *by
allowing it to drink as much of the cow's milk as it feels
inclined for,* let the quantity usually allotted for a meal
be mixed, consisting, for *the first week,* of three parts
milk and one part hay-tea. *The only nourishing infusion of hay is that which is made from the best and sweetest hay, cut by a chaff-cutter into pieces about two inches
long,* and put into an earthen vessel; over this boiling
water should be poured, and the whole allowed to stand
for two hours, during which time it ought to be kept
carefully closed. After the first week, the proportions
of milk and hay-tea may be equal; then composed of
two thirds of hay-tea and one of milk; and at length one
fourth part of milk will be sufficient. This food should
be given to the calf in a lukewarm state *at least three,
if not four times a day, in quantities averaging three
quarts at each meal,* but gradually increasing to four
quarts *as the calf grows older. Towards the end of the
second month,* beside the usual quantity given at each
meal (composed of three parts of the infusion and one
of milk), a small wisp or bundle of hay is to be laid
before the calf, which will gradually come to eat it; but,

if the weather is favorable, as in the month of May, the beast may be turned out to graze in a fine, sweet pasture, well sheltered from the wind and sun. This diet may be continued until towards the latter end of the third month, when, if the calf grazes heartily, each meal may be reduced to less than a quart of milk, with hay-water; or skimmed milk or fresh butter-milk may be substituted for new milk. At the expiration of the third month the animal will hardly require to be fed by hand, though, if this should still be necessary, one quart of the infusion given daily, and which during the summer need not be warmed, will be sufficient." The hay-tea should be made fresh every two days, as it soon loses its nutritious quality.

This and other preparations are given not because they are better than milk, than which nothing is better adapted to fatten a calf, or promote its growth, but simply to economize by providing the most suitable and cheaper substitutes. Experience shows that the first two or three calves are smaller than those that follow; and hence, unless they are pure-bred, and to be kept for the blood, they are not generally thought to be so desirable to raise for the dairy as the third or fourth, and those that come after, up to the age of nine or ten years. On this point opinions differ.

According to the comparative experiments of a German agriculturist, cows which as calves had been allowed to suckle their dams from two to four weeks brought calves which weighed only from thirty-five to forty-eight pounds; while others, which, as calves, had been allowed to suckle from five to eight weeks, brought calves weighing from sixty to eighty pounds. It is difficult to see how there can be so great a difference, if, indeed, there is any; but it may be worthy of careful observation and experiment, and as such it is stated in this connection. The increased size of the calf would be due to the larger size to which the cow would attain; and if as a calf she were allowed to run with her dam in the pasture four or five months, taking all the milk she wanted, she would doubtless be kept growing on in a thriving condition. But taking a calf from the cow at four or even eight weeks must check its growth to some extent, and this may be avoided by feeding liberally, and bringing up by hand.

After the calf is fully weaned, there is nothing very peculiar in the general management. A young animal will require for the first few months — say up to the age of six months — an average of five or six pounds daily of good hay, or its equivalent. At the age of six months it will require from four and a half to five pounds, and at the end of the year from three and a half to four pounds of good hay, or its equivalent, for every one hundred pounds of its live weight; or, in other words, about three and a half or four per cent. of its live weight. At two years old it will require three and a half, and some months later three per cent. of its live weight daily in good hay or its equivalent. Indian-corn fodder, either green or cured, forms an excellent and wholesome food at this age.

The heifer should not be pampered, nor yet poorly fed or half starved, so as to receive a check in her growth. An abundant supply of good healthy dairy food and drink will do all that is necessary up to the time of having her first calf, which should not ordinarily be till the age of three years, though some choose to allow them to come in at two or a little over, on the ground that it early stimulâtes the secretion of milk, and that this will increase the milking propensity through life. This is undoubtedly the case, as a general rule; but I think greater injury is done by checking

the growth, unless the heifer has been fed up to large
size and full development from the start, in which case
she may perhaps take the bull at fifteen or eighteen
months without injury.   I have had several come in as
early as two years, and one at less than twenty months.
This last was not by design, however, and I would
rather have given a considerable sum than had it happen, as she was an exceedingly beautiful pure-bred Jersey, and I was desirous to have her attain to good size
and growth.   Even if a heifer comes in at two years, it
is generally thought desirable to let her run farrow for
the following year, which will promote her growth and
more perfect development.

The feeding which young stock often get is not such
as is calculated to make good-sized or valuable cattle of
them.   They are often fed on the poorest of hay or
straw through the winter, not unfrequently left exposed
to cold, unprotected and unhoused, and thus stinted in
their growth.   This seems to me to be the very worst
economy, or rather no economy at all.   Properly viewed,
it is an extravagant wastefulness which no farmer can
afford.   No animal develops its good points under such
treatment; and if the starving system is to be followed
at all, it had better be after the age of two or three
years, when the animal's constitution has attained
strength and vigor to resist ill treatment.

To raise up first-rate milkers, it is absolutely necessary to feed on dairy food even while young.   No
matter how fine the breed is, if the calf is raised on
poor, short feed, it will never be so good a milker as if
raised on better keeping; and hence, in dairy districts, where calves are raised at all, they ought to be
allowed the best pasture during the summer, and good
sweet and wholesome food during the winter.

CHAPTER VI.

CULTURE OF GRASSES AND OTHER PLANTS REC-
OMMENDED FOR FODDER.

As already stated, the grasses in summer, and hay in winter, form the most natural and important food for milch cows ; and, whatever other crops come in as ad-ditional, these will form the basis of all systems of feeding.

The nutritive qualities of the grasses differ widely ; and their value as feed for cows will depend, to a con-siderable extent, on the management of pastures and mowing-lands.

If the turf of an old pasture is carefully examined, it will be found to contain a large variety of grasses and plants adapted for forage ; some of them valuable for one purpose, and some for another. Some of them, though possessing a lower percentage of nutritive constituents than others, are particularly esteemed for an early and luxuriant growth, furnishing a sweet feed in early spring, before other grasses appear ; some of them, for starting more rapidly than others, after being eaten off by cattle, and consequently of great value as pasture grasses. Most grasses will be found to be of a social character, and to do best in a large mixture with other varieties.

In forming a mixture for pasture grasses, the pecu-liarities of each species should, therefore, be regarded :

15

as the time of flowering, the habits of growth, the soil and location on which it grows best, and other characteristics. Among the grasses found on cultivated lands, in this country, the following are considered as among the most valuable for ordinary farm cultivation; some of them adapted to pastures, and others almost exclusively to mowing and the hay crop: Timothy (*Phleum pratense*). Meadow Foxtail (*Alopecurus pratensis*). June, or Kentucky Blue Grass (*Poa pratensis*). Fowl meadow (*Poa serotina*). Rough-stalked Meadow (*Poa trivialis*). Orchard Grass (*Dactylis glomerata*). Perennial Rye Grass (*Lolium perenne*). Italian Rye Grass (*Lolium italicum*). Redtop (*Agrostis vulgaris*). English Bent (*Agrostis alba*). Meadow Fescue (*Festuca pratensis*). Tall Oat Grass (*Arrhenatherum avenaceum*). Sweet-scented Vernal (*Anthoxanthemum odoratum*). Hungarian Grass (*Panicum Germanicum*). Red Clover (*Trifolium pratense*). White or Dutch Clover (*Trifolium repens*), and some others.

Of these, the most valuable, all things considered, is the first, or Timothy (Fig. 56). It forms a large proportion of what is commonly called English, or in some sections meadow hay, though it originated and was first cultivated in this country. It contains a large percentage of nutritive matter, in comparison with other agricultural grasses. It thrives best on moist, peaty, or loamy soils, of medium tenacity, and is not well suited to very light, sandy lands. On very moist soils its root is almost always fibrous ; while on dry and loamy ones it is bulbous. On soils of the former description, which it especially affects, its growth is rapid, and its yield of hay large, sometimes amounting to three and four tons to the acre, depending much, of course, on cultivation. But, though very valuable for hay, it is not adapted to pastures, as it will neither endure severe grazing, nor

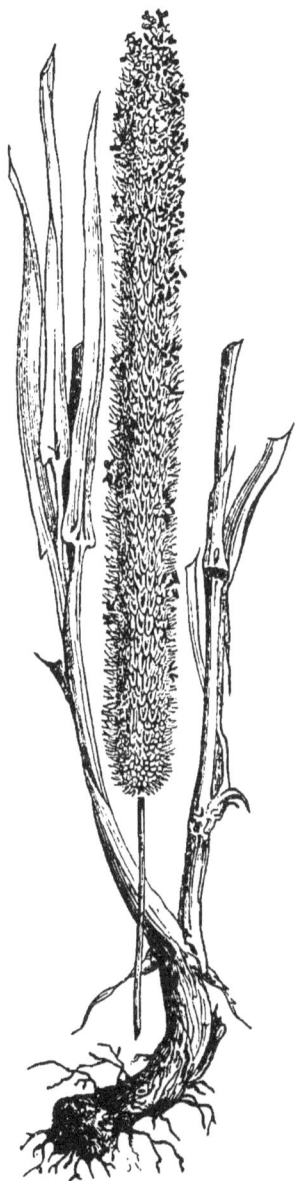

Fig. 56. Timothy grass.

Fig 57. June grass.

is its aftermath to be compared with meadow foxtail
and some of the other grasses.

JUNE GRASS (Fig. 57), better known in some sections as Kentucky Blue grass, is very common in most sections of the country, especially on limestone lands, forming a large part of the turf, wherever it flourishes, and being universally esteemed as a pasture grass. It starts early, but varies much in size and appearance, according to the soil; growing in some places with the utmost luxuriance, and forming the predominant grass; in others, yielding to the other species. If cut at the time of flowering, or a few days after, it makes a good and nutritive hay, though it is surpassed in nutritive quali ties by several of the other grasses. It starts slowly after being cut, especially if not cut very early. But its herbage is fine and uniform, and admirably adapted to lawns, growing well in almost all soils, though it does not endure very severe droughts. It withstands, however, the frosts of winter better than most other grasses.

In Kentucky, a section where it attains its highest perfection and luxuriance, ripening its seed about the 10th of June, and in latitudes south of that, it sometimes continues green through the mild winters. It requires three or four years to become well set, after sowing, and it does not attain its highest yield as a pasture grass till the sod is even older than that. It is not, therefore, suited to alternate husbandry, where land usually remains in grass but two or three years before being ploughed up. In Kentucky it is sown any time in winter when the snow is on the ground, three or four quarts of seed being used to the acre. In spring the seeds germinate, when the sprouts are exceedingly fine and delicate. Stock is not allowed on it the first year.

The MEADOW FOXTAIL (Fig. 58) is also an excellent pasture grass. It somewhat resembles Timothy, but is ear lier, has a softer spike, and thrives on all soils except the

Fig. 58. Meadow Foxtail.   Fig. 59. Orchard grass.

dryest.   Its growth is rapid, and it is greatly relished by
stock of all kinds.· Its stalk and leaves are too few and
light for a field crop, and it shrinks too much in curing to

15*

be valuable for hay. It flourishes best in a rich, moist, and rather strong soil, sending up a luxuriant aftermath when cut or grazed off, which is much more valuable, both in quantity and nutritive value, than the first crop. In all lands designed for permanent pasture, therefore, it should form a considerable part of a mixture. It will endure almost any amount of forcing, by liquid manures, or irrigation. It requires three or four years, after sowing, to gain a firm footing in the soil. The seed is covered with the soft and woolly husks of the flower, and is consequently light; weighing but five pounds to the bushel, and containing seventy-six thousand seeds to the ounce.

The ORCHARD GRASS, or ROUGH COCKSFOOT (Fig. 59), for pastures, stands preëminent. This is a native of this country, and was introduced into England, from Virginia, in 1764, since which time its cultivation has extended into every country of Europe, where it is universally held in very high estimation. The fact of its being very palatable to stock of all kinds, its rapidity of growth, and the luxuriance of its aftermath, with its power of enduring the cropping of cattle, have given it a very high reputation, especially as a pasture grass. It blossoms earlier than Timothy; when green is equally relished by milch cows; requires to be fed closer, to prevent its forming tufts and growing up to seed, when it becomes hard and wiry, and loses much of its nutritive quality. As it blossoms about the same time, it forms an admirable mixture with red clover, either for permanent pasture or mowing. It resists drought, and is less exhausting to the soil than either rye grass or Timothy. The seed weighs twelve pounds to the bushel, and when sown alone requires about two bushels to the acre.

The ROUGH-STALKED MEADOW GRASS (Fig. 60) is somewhat less common than June grass, but is considered as

Fig. 60. Rough-stalked Meadow grass.   Fig. 61. Rye grass.

equally valuable.. It grows best on moist, sheltered meadows, where it flowers in June and July. It is easily dis-

tinguished from June grass, by having a rough sheath, while the latter has a smooth one, and by having a fibrous root, while the root of June grass is creeping. It possesses very considerable nutritive qualities, and comes to perfection at a desirable time; is exceedingly relished by cattle, horses, and sheep. For suitable soils it should form a portion of a mixture of seeds, producing, in mixture with other grasses which serve to shelter it, a large yield of hay, far above the average of grass usually grown on a similar soil. It should be cut when the seed is formed. Seven pounds of seed to the acre will produce a good sward. The grass loses about seventy per cent. of its weight in drying. The nutritive qualities of its aftermath exceed very considerably those of the crop cut in the flower or in the seed.

FOWL MEADOW GRASS is another indigenous species, of great value for low and marshy grounds, where it flourishes best; and, if cut and properly cured, makes a sweet and nutritious hay, which, from its fineness, is eaten by cows without waste. According to Sinclair, who experimented, with the aid of Sir Humphrey Davy, to ascertain its comparative nutritive properties, it is superior, in this respect, to either meadow foxtail, orchard grass, or tall meadow oat grass; but it is probable that he somewhat overrates it. If allowed to stand till nearly ripe, it falls down, but sends up innumerable flowering stems from the joints, so that it continues green and luxuriant till late in the season. It thrives best in mixture with other grasses, and deserves a prominent place in all mixtures for rich, moist pastures, and low mowing-lands.

RYE GRASS (Fig. 61) has a far higher reputation abroad than in this country, and probably with reason; for it is better adapted to a wet and uncertain climate than to a dry and hot one. It varies exceedingly, depending much on soil and culture ; but, when cut in the blossom to make into hay, it possesses very considerable nutritive power. If allowed to get too ripe, it is hard and wiry, and not relished by cows. The change from a juicy and nutritious plant to woody fibre, possessing but little soluble matter, is very rapid. Properly managed, however, it is a tolerably good grass, though not to be compared to Timothy, or orchard grass.

ITALIAN RYE GRASS (Fig. 62) has also been cultivated to considerable extent in this country, but with less satisfactory results than are obtained from it in Europe, where it endures all climates, giving better crops, both in quantity and quality, than the perennial rye grass. It is one of the greatest gluttons of all the grasses, and luxuriates in frequent irrigation with liquid manure, though it is said to stand the drought very well. The soils best adapted to it are rich, moist, and fertile, of medium tenacity; and it is admirably adapted to the purposes of soiling, as it endures repeated cutting, rapidly sending up luxuriant crops. For rich soils near the barn, used for the growth of crops for soiling, therefore, it may be confidently used as a profitable addition to our list of cultivated grasses.

REDTOP (Fig. 63) is a grass familiar to every farmer in the country. It is the Herd's grass of Pennsylvania, while in New York and New England it is known by a great variety of names, and assumes a great variety of forms, according to the soil in which it grows. It is well adapted to almost every soil, though it seems to prefer a moist loam. It makes a profitable crop for spending, in the form of hay, though its yield is less than that of Timothy. It is well suited to our permanent pastures, where it should be fed close, otherwise it becomes wiry and innutritious, and cattle refuse it. It stands

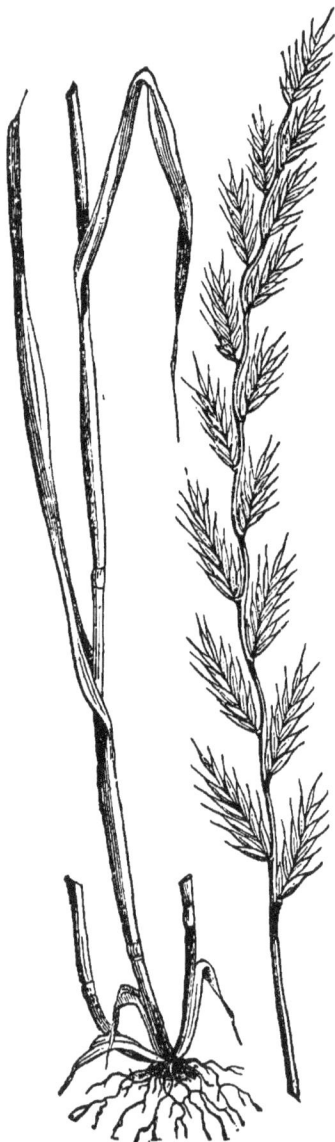

Fig. 62. Italian Rye grass.   Fig. 63. Redtop.

the climate of the country as well as any other grass,
and so forms a valuable part of any mixture for

Fig. 64.   English Bent.   Fig. 65.   Meadow Fescue.

pastures and permanent mowing-lands; but it is probably rather overrated by us.

ENGLISH BENT (Fig. 64), known also by a great variety of other names, is also largely cultivated in some sections. It closely resembles redtop, but may be distinguished from it by the roughness of the sheaths when the hand is drawn from above downwards. It possesses much the same qualities as redtop.

MEADOW FESCUE (Fig. 65) is one of the most common of the fescue grasses, and is said to be the Randall grass of Virginia. It is an excellent pasture grass, forming a very considerable portion of the turf of old pastures and fields; and is more extensively propagated and diffused by the fact that it ripens its seeds before most other grasses are cut, and sheds them to spring up and cover the ground. Its long and tender leaves are much relished by cattle. It is rarely sown in this country, notwithstanding its great and acknowledged value as a pasture grass. If sown at all, it should be in mixture with other grasses, as orchard grass, rye grass, or June grass. It is of much greater value at the time of flowering than when the seed is ripe.

The TALL OAT GRASS (Fig. 66) is the Ray grass of France. It furnishes a luxuriant supply of foliage, is valuable either for hay or for pasture, and has been especially recommended for soiling purposes, on account of its early and luxuriant growth. It is often found on the borders of fields and hedges, woods and pastures, and is sometimes very plenty in mowing-lands. After being mown it shoots up a very thick aftermath, and on this account, partly, is regarded as nearly equal for excellence to the common foxtail.

It grows spontaneously on deep, sandy soils, when once naturalized. It has been cultivated to a considerable extent in this country, and is esteemed by those who know it mainly for its early, rapid, and late growth,

Fig. 66. Tall Oat grass.

Fig. 67. Sweet-scented Vernal.

making it very well calculated as a permanent pasture grass. It will succeed on tenacious clover soils.

16

The SWEET-SCENTED VERNAL GRASS (Fig. 67) is one of the earliest in spring and one of the latest in autumn; and this habit of growth is one of its chief excellences, as it is neither a nutritious grass nor very palatable to stock of any kind, nor does it yield a very good crop. It is very common all over New England and the Middle States, coming into old worn-out fields and moist pastures spontaneously, and along every roadside. It derives its name from its sweetness of smell when partially wilted, or crushed in the hand, and it is this chiefly that gives the delicious fragrance to all new-mown hay. It is almost the only grass that possesses a strongly-marked aromatic odor, which is imparted to other grasses with which it is cured. Its seed weighs eight pounds to the bushel. In mixtures for permanent pastures it may be of some value.

HUNGARIAN GRASS, or Millet (Fig. 68), is an annual forage plant, introduced into France in 1815, and more recently into this country. It germinates readily and withstands the drought remarkably, remaining green when other grasses are parched and dried up. It has numerous succulent leaves, which furnish an abundance of sweet fodder, greatly relished by stock of all kinds. It attains its greatest luxuriance on soils of medium consistency and richness, but does very well on light and dry plains.

RED CLOVER (Fig. 69) is an artificial grass of the leguminous family, and one of the most valuable of cultivated plants for feeding to dairy cows. It flourishes best on tenacious soils and stiff loams. Its growth is rapid, and a few months after sowing are sufficient to supply an abundant sweet and nutritious food. In the climate of New England clover should be sown in the spring of the year, while most of the natural grasses do far better sown in the fall. It is often sown with per-

Fig. 69.　Red Clover.　　　　Fig 68.　Hungarian grass.

fect success on the late snows of March or April, and
soon finds its way down into the soil and takes a vigor·

ous root. It is valuable not only as a forage plant, but as shading the ground, and thereby increasing its fertility.

The introduction of clover among the cultivated plants of the farm has done more, perhaps, for modern agriculture than that of any other single plant. It has now come to be considered indispensable in all good dairy districts.

Fig. 70. White Clover.

WHITE CLOVER (Fig. 70), often called Honeysuckle, is also widely diffused over this country, to which it is undoubtedly indigenous. As a mixture in all pasture grasses it holds a very high rank, as it is exceedingly sweet and nutritious, and relished by stock of all kinds. It grows most luxuriantly in moist grounds and moist seasons, but easily accommodates itself to a great variety of circumstances.

With respect to the mixtures of grass-seeds most profitable for the dairy farmer, no universal rule can be given, as they depend very much upon the nature of the soil and the locality. The most important point to be observed, and one in which we, as a body, are perhaps most deficient, is to use a large number of species, with smaller quantities of each than those most commonly used. This is nature's rule; for, in examin-

ing the turf of a rich old pasture, we shall find a large
number of different species growing together, while, if
we examine the turf of a field sown with only one or
two different species, we find a far less number of plants
to the square foot, even after the sod is fairly set. No
improvement in grass culture is more important, it
seems to me. I have suggested, in another place, a
large number of mixtures adapted to the different
varieties of soil and circumstance, together with the
reasons for the mixture in many instances. (See *A
Practical Treatise on Grasses and Forage Plants*, comprising their Natural History, Comparative Nutritive
Value, Methods of Cultivating, Cutting, and Curing, and
the Management of Grass Lands, &c. 400 pp. 8vo., with
illustrations.) As an instance of what I should consider
an improvement on our ordinary mixtures for *permanent
pastures*, I would suggest the following as likely to
give satisfactory results, dependent, of course, to a considerable extent, on the nature and preparation of the
soil:

Meadow Foxtail, flowering in May and June,	. . . .	2 pounds
Orchard Grass,	" " " " "	. . . . 6 "
Sweet-scented Vernal, "	" April and May,	. . . . 1 "
Meadow Fescue, "	" May and June,	. . . . 2 "
Redtop, "	" June and July,	. . . . 2 "
June Grass, "	" May and June,	. . . . 4 "
Italian Rye Grass, "	" June,	. . . . . 4 "
Perennial Rye Grass, "	" June,	. . . . . . 6 "
Timothy, "	" June and July,	. . . 3 "
Rough-stalked Meadow Grass, flowering in June and July,		2 "
Perennial Clover, flowering in June,	. . . .	3 "
White Clover, "	" May to September,	. . 5–40 "

For mowing-lands the mixture would, of course, be
somewhat changed. The meadow foxtail and sweet-
scented vernal would be left out entirely, and some six
or eight pounds added to the Timothy and red clover.

16*

The proper time to lay down lands to grass in the lati-
tude of New England is August or September, and no
grain crop should be sown with the seed.

Stiff or clayey pastures should never be over-
stocked, but when fed pretty close the grasses are
far sweeter and more nutritious than when they are
allowed to grow up rank and coarse; and if, by a want
of sufficient feeding, they get the start of the stock, and
grow into rank tufts, they should be cut and removed,
when a fresh grass will start up, similar to the aftermath of mowing-lands, which will be greedily eaten.
Grasses for curing into hay should be cut either at the
time of flowering or just before, especially if designed
for milch cows. They are then more succulent and
juicy, and, if properly cured, form the sweetest food.

Grass cut in the blossom will make more milk than if
allowed to stand later. Cut a little before the blossom-
ing, it will make more than when in the blossom; and
the cows prefer it, which is by no means an unimportant
consideration, since their tastes should always be con-
sulted. Grass cut somewhat green, and properly cured,
is next to fresh, green grass in palatable and nutritive
qualities. And so a sensible practical farmer writes
me: "The time of cutting grass depends very much
upon the use you wish to make of it. If for working
oxen and horses, I would let it stand till a little out of
the blossom; but if to feed out to new milch cows in
the winter, I would prefer to cut it very green. It is
then worth for the making of milk in the winter almost
double that cut later." Every farmer knows the milk-
producing properties of rowen, which is generally cut
before it blossoms.

No operation on the farm is of greater importance
to the dairyman than the cutting of his grass and
the manner of curing hay, and in this respect the

practice over the country generally is susceptible of very great improvement. The chief object is to preserve the sweetness and succulence of grass in its natural state, so far as it is possible; and this object cannot be gained by exposing it too long to the scorching suns and the drenching rains to which we are liable in this climate. We generally try to make our hay too much.

As to the best modes of curing clover, my own experience and observation accord with that of several practical farmers, who write me as follows: "My method of curing clover is this: What is mown in the morning I leave in the swath, to be turned over early in the afternoon. At about four o'clock, or while it is still warm, I put it into small cocks with a fork, and, if the weather is favorable, it may be housed on the fourth or fifth day, the cocks being turned over on the morning of the day it is to be carted. By so doing all the heads and leaves are saved, and these are worth more than the stems. This has been my method for the last ten years. For new milch cows in the winter I think there is nothing better. It will make them give as great a flow of milk as any hay, unless it be good rowen." Another says: "When the weather bids fair to be good, I mow it after the dew is off, and cock it up after being wilted, using the fork instead of rolling with the rake, and let it remain several days, when it is fit to put into the barn." And another: "I mow my clover in the forenoon, and towards night of the same day I take forks and pitch it into cocks and let it stand till it cures. The day I cart it, I turn the cocks over, so as to air the lower part. I then put it into the mow with all the leaves and heads on, and it is as nice and green as green tea. I think it worth for milch cows and sheep as much per ton as English hay." And still another: "I have found no better hay for farm stock than good clover, cut in season. For milch cows it is much better than Timothy. The rowen crop is bettei than any other for calves."

INDIAN CORN makes an exceedingly valuable fodder, both as a means of carrying a herd of milch cows through our severe droughts of summer, and as an article for soiling cows kept in the stall. No dairy farmer will neglect to sow an extent in proportion to the number of cows he keeps. The most common practice is to sow in drills from two and a half to three feet apart, on land well tilled and thoroughly manured, making the drills from six to ten inches wide with the plough, manuring in the furrow, dropping the corn about two inches apart, and covering with the hoe. In this mode of culture the cultivator may be used between the rows when the corn is from six to twelve inches high, and unless the ground is very weedy no other after culture is generally needed. The first sowing usually takes place about the 20th of May, and this is succeeded by other sowings at. intervals of a week or ten days, till July, in order to have a succession of green fodder. But, if it is designed to cut it up to cure for winter use, an early sowing is generally preferred, in order to be able to cure it in warm weather, in August or early in September. Sown in this way, about three or four bushels of corn are required for an acre, since, if sown thickly, the fodder is better, the stalks smaller, and the waste less.

The chief difficulty in curing corn cultivated for this purpose, and after the methods spoken of, arises mainly from the fact that it comes at a season when the weather is often colder, the days shorter, and the dews heavier, than when the curing of hay takes place. Nor is the curing of corn cut up green so easy and simple as that of drying the stalks of Indian corn cut above the ear, as in our common practice of topping. The plant is then riper, less juicy, and cures more readily. The method sometimes adopted is to cut and tie into small bundles, after it is somewhat wilted, and stook upon the ground, where it is allowed to stand, subject to all the changes of the weather, with only the protection of the stook itself. The stooks consist of bunches of stalks first bound in small bundles, and are made sufficiently large to prevent the wind from blowing them over. The arms are thrown around the tops to bring them together as closely as possible, when the tops are broken over or twisted together, or otherwise fastened, in order to make the stook " shed the rain " as well as possible. In this condition they stand out till sufficiently dried to put into the barn. Corn fodder is very excellent for young dairy stock.

COMMON MILLET (*Panicum miliaceum*) is another very valuable crop for fodder in soiling, or to cure for winter use, but especially to feed out during our usual periods of drought. Many varieties of millet are cultivated in this country, the ground being prepared and treated as for oats. If designed to cut for green fodder, half a bushel of seed to the acre should be used, if to ripen seed, twelve quarts, sown broad-cast, about the last of May or early in June. A moist loam or muck is the best adapted to millet; but I have seen very great crops grown on dry upland. It is very palatable and nutritious for milch cows, both green and when properly cured. The curing should be very much like clover, care being taken not to over-dry it. For fodder, either green or cured, it is cut before ripening. In this state all cattle eat it as readily as green corn, and a less extent will feed them. Millet is worthy of a widely-extended cultivation, particularly on dairy farms. Indian millet (*Sorghum vulgare*) is another cultivated variety.

RYE, as a fodder plant, is chiefly valuable for its early growth in spring. It is usually sown in September or October, from the middle to the end of September being, perhaps, the most desirable time, on land previously cultivated and in good condition. If designed to ripen only, a bushel of seed is required to the acre, evenly sown; but, if intended for early fodder in spring, two or two and a half bushels per acre of seed should be used. On warm land the rye can be cut green the last of April or first of May; and care should be taken to cut early, as, if allowed to advance too far towards maturity, the stalk becomes hard and unpalatable to cows.

OATS are also sometimes used for soiling, or for feeding green, to eke out a scanty supply of pasture feed; and for this purpose they are valuable. They should be sown on well-tilled and well-manured land, about four bushels to the acre, towards the last of April or first of May. If the whole crop is to be used as green fodder, five bushels of seed will not be too much on strong, good soil. They will be sufficiently grown to cut by the first of July, or in some sections earlier, depending on location.

The CHINESE SUGAR-CANE also may deserve attention as a fodder plant. Experiments hitherto made seem to show that when properly cultivated, and cut at the right time, it is a palatable and nutritious plant, while many of the failures have been the result of too early cutting. For a fodder crop the drill culture is preferable, both on account of the larger yield obtained and to prevent it from becoming too hard and stalky.

THE POTATO (*Solanum tuberosum*) is the first of the root crops to be mentioned. This produces a large quantity of milk, though the quality is inferior. The market value of this root is, at the present time, too great to allow of feeding extensively with it, even in milk-dairies, where it is most valuable as food for cows; still, there are locations where it may be judicious to cultivate this root for dairy feed, and in all circum-stances there is a certain portion of the crop of un marketable size, which will be of value fed to milch cows or swine. It should be planted in April or May, but in many sections in June, on good mellow soil, first thoroughly ploughed and harrowed, then furrowed three feet apart, and manured in the furrows with a mixture of ashes, plaster of Paris, and salt. The seed may be dropped in the furrows, one foot apart, after the drill system, or in hills, two and a half or three feet apart, to be covered with the plough by simply turning the furrows back, after which the whole should be rolled with the field roller, where it can be done.

If the land is not already in good heart from continued cultivation, a few loads of barn-yard manure may be spread, and ploughed under by the first ploughing. Used in this way, it is far less liable to cause the rot than when put in the hill. If a sufficient quantity of wood-ashes is not at hand, sifted coal-ashes will answer the purpose, and are said to be valuable as a preventive of the rot. In this way one man, two boys, and a horse, can plant from three to four acres a day on mellow land. I have planted two acres a day on the sod, the manure being first spread on the grass, a furrow made by a yoke of oxen and one man, another following after and dropping, a foot apart, along the outer edge of the furrow on the grass. By quick work, one hand can nearly keep up with the plough in dropping. When arrived at the end of the piece, a back furrow is turned up to the

potatoes, and a good ploughman will cover nearly all
without difficulty. On the return-furrow the man or
boy who dropped follows after, covering up any that
may be left or displaced, and smoothing off the top
of the back-furrows where necessary. Potatoes thus
planted came out as fine as I ever saw any.

The cost of cultivation in this mode, it must be
evident is but trifling compared with the slower
method of hand-planting. The plan will require a skilful ploughman, a quick, active lad, and a good yoke
of oxen, and the extent of the work will depend
somewhat on the state of the turf. The nutritive
equivalent in potatoes for one hundred pounds of good
hay is 3.19 pounds; that is, it will take 3.19 pounds of
potatoes to afford the same amount of nourishment
as one pound of hay. The great value of roots is as a
change or condiment, calculated to keep the animal in
a healthy condition.

THE CARROT (*Daucus carota*) is somewhat extensively fed, and is a valuable root for milch cows. This,
like the potato, has been cultivated and improved from
a wild plant. Carrots require a deep, warm, mellow
soil, thoroughly cultivated, but clean and free from
weed-seed. The difference between a very good profit
and a loss on the crop depends much on the use of
land and manures perfectly free from foul seeds of
any kind. Ashes, guano, sea-weed, ground bone, and
other similar substances, or thoroughly-rotted and
fermented compost, will answer the purpose.

After ploughing deep, and harrowing carefully, the
seed should be sown with a seed-sower, in drills about
eighteen inches apart, at the rate of four pounds to
the acre, about the middle or twentieth of April. The
difference between sowing by the first of May and the
tenth of June in New England is said to be nearly one

third in the crop on an average of years. In weeding, a little wheel-hoe is invaluable, as with it a large part of the labor of cultivation is saved. A skilful hand can run this hoe within half an inch of the young plants without injury, and go over a large space in the course of a day, if the land was properly prepared in the first place.

The American farmer should always plan to economize labor. That is the great item of expense on the farm. I do not mean that he should try to shirk or avoid work, but that he should make the least amount of work accomplish the largest and most profitable results. Labor-saving machinery on the farm is applied not to reduce the number of hours' labor, or to make the owner a man of leisure,—who is, generally, the unhappiest man in the world,—but to enable him to accomplish the greatest results in the same time that he would be compelled to labor to obtain smaller ones.

Carrots will continue to grow and increase in size late into the fall. When ready to dig, plough around as near to the outside rows as possible, turning the furrow away from the row. Then take out the carrots, pulling off the tops, and throw the carrots and tops into separate heaps on the ploughed furrows. In this way a man and two boys can harvest and put into the cellar over a hundred bushels a day.

The TURNIP (*Brassica rapa*) and the Swedish turnip or ruta baga (*Brassica campestris*) are also largely cultivated as a field crop to feed to stock; and for this purpose numberless varieties are used, furnishing a great amount of succulent and nutritious food, late into winter, and, if well kept, late into spring. The chief objection to the turnip is that it taints the milk. This may be remedied, to a considerable extent, if not wholly, by the use of salt, or salt hay, and by feeding at the time of milking, or immediately after, or by steaming before feeding, or putting a small quantity of the solution of nitre into the pail, and milking upon it.

Turnips may be sown any time in June, in rich land, well mellowed by cultivation. Very large crops are often obtained sown as late as the middle of July, or first of August, on an inverted sod. The Michigan or double-mould-board plough leaves the land light, and in admirable condition to harrow, and drill in turnips. A successful root-grower last year cut two tons of hay to the acre, on the 23d of June, and after it was removed from the land spread eight cords of rotten kelp to the acre, and ploughed in; after which about three cords of fine old compost manure were used to the acre, which was sown with ruta baga seed, in drills, three feet apart, plants thinned to eight or ten inches in the drill. No after cultivation was required. On the 15th of November he harvested three hundred and seventy bushels of splendid roots to the acre, carefully measured off.

The nutritive equivalent of Swedish turnips as compared with good meadow hay is 676, taking hay as a standard at 100; that is, it would require 6.76 lbs. of turnips to furnish the same nutriment as one pound of good hay; but, fed in connection with other food, as hay, for instance, perhaps five pounds of turnips would be about equal to one pound of hay.

The English or round turnip is usually sown broadcast after some other crop, and large and valuable returns are often obtained. The Swede is sown in drills. Both these varieties are used for the production of milk.

The chief objection to the turnip crop is that it leaves many kinds of soil unfit for a succession of some other crops, like Indian corn, for instance. In some sections no amount of manuring appears to make corn do well after turnips or ruta bagas.

The MANGOLD WURZEL, a variety of the *Beta vulgaris*, is often cultivated with great success in this country, and fed to cows with advantage, furnishing a succulent and nutritive food in winter and spring. The crop is somewhat uncertain. When it does well an enormous yield is often obtained; but it often proves a failure, and is not, on the whole, quite as reliable as the ruta baga, though a more valuable crop when the yield is good. It is cultivated like the common beet, in moist, rich soils, three pounds of seed to the acre. The leaves may be stripped off, towards fall, and fed out, without injury to the growth of the root. Both mangolds and turnips should be cut with a root-cutter, before being fed out.

The PARSNIP (*Pastinaca sativa*) is a very sweet and nutritive article of fodder, and adds richness and flavor to the milk. It is worthy of extended culture in all parts of this country where dairy husbandry is pursued. It is a biennial, easily raised on deep, rich, well-cultivated and well-manured soils, often yielding enormous crops, and possessing the advantage of withstanding the severest winters. As an article of spring feeding, therefore, it is exceedingly valuable. Sown in April or May, it attains a large growth before winter. Then, if desirable, a part of the crop may be harvested for winter use, and the remainder left in the ground till the frost is out, in March or April, when they can be dug as wanted, and are exceedingly relished by milch cows, and stock of all kinds. They make an admirable feed at the time of milking, and produce the richest cream, and the yellowest and finest-flavored butter, of any root with which I am acquainted. The good dairy farmers on the island of Jersey often feed to their cows from thirty to thirty-five pounds of parsnips a day, in addition to hay or grass.

Both practical experiment and scientific analysis prove this root to be eminently adapted to dairy stock, where the richness of milk or fine-flavored butter is any object. For mere milk-dairies, it is not quite so valuable, probably, as the Swedish turnip. The culture is similar to that of carrots, a rich, mellow, and deep loam being best; while it has a great advantage over the carrot in being more hardy, and rather less liable to injury from insects, and more nutritive. For feeding and fattening stock it is eminently adapted.

To be sure of a crop, fresh seed must be had, as it cannot be depended on more than one year. For this reason, the largest and straightest roots should be allowed to stand for seed, which, as soon as nearly ripe, should be taken off and spread out to dry, and carefully kept for use. For field culture the hollow-crowned parsnip is the best and most profitable ; but on thin, shallow soils the turnip-rooted variety should be used. Parsnips may be harvested like carrots, by ploughing along the rows. Let butter or cheese dairymen give this crop a fair and full trial, and watch its effect on the quality of the milk and butter.

The KOHL RABI (*Brassica oleracea*, var. *caulorapa*) is also cultivated to a considerable extent in this country, to feed to stock. It is supposed to be a hybrid between the cabbage and the turnip, and is often called the cabbage-turnip, having the root of the former, with a turnip-like or bulbous stem. The special reason for its more extensive cultivation among us is its wonderful indifference to droughts, in which it seems to flourish best, and to bring forth the most luxuriant crops. It also withstands the frosts remarkably, being a hardy plant. It yields a somewhat richer quality of milk than the ordinary turnip, and the crop is generally admitted to be as abundant and profitable. I have seen very large crops of it produced by the ordinary turnip or cabbage cultivation. As in cabbage culture, it is best to sow the seed in March or April, in a warm and well-enriched seed-bed; from which it is transplanted in May, and set out after the manner of cabbages in garden culture. It bears transplanting better than most other roots. Insects injure it less than the turnip, dry weather favors it, and it keeps well through the winter. For these reasons, it must be regarded as a valuable addition to our list of forage plants adapted to dairy farming. It grows well on stronger soils than the turnip.

LINSEED MEAL is the ground cake of flax-seed, after the oil is pressed out. It is very rich in fat-forming principles, and given to milch cows it increases the quality of butter, and keeps them in condition. Four or five pounds a day are sufficient for cows in milk, and this amount will effect a great saving in the cost of other food, and at the same time make a very rich milk. It is extensively manufactured in this country, and largely exported, but is worthy of more general use here. It must not be fed in too large quantities to milch cows, for it would be liable to give too great a tendency to fat, and thus affect the quantity of milk.

RAPE-CAKE possesses much the same qualities. It is the residuum after pressing the oil from rape-seed.

COTTON-SEED MEAL is an article of comparatively recent introduction. It is obtained by pressing the seed of the cotton-plant, which extracts the oil, when the cake is crushed or ground into meal, which has been found to be a very valuable article for feeding stock. An analysis has been given on a preceding page, which shows it to be equal or superior to linseed meal. Practical experiments are needed to establish it. It is pre-

17*

pared chiefly in Providence, R. I., and is for sale in the market at a very reasonable price.

The MANURES used in this country in the culture of the plants mentioned above are mostly such as are made on the farm, consisting chiefly of barn-yard composts of various kinds, with often a large admixture of peat-mud. There are few farms that do not contain substances which, if properly husbanded, would add very greatly to the amount of manure ordinarily made. The best of the concentrated manures, which it is sometimes necessary to use, for want of time and labor to prepare enough on the farm, is, unquestionably, Peruvian guano. The results of this, when properly applied, are well known and reliable, which can hardly be said of any other artificial manure offered for the farmer's notice. The chief objection to depending on manures made off the farm is, in the first place, their great expense; and in the second, which is equally important, the fact that, though they may be made valuable, and produce at one time the best results, a want of care in the manufacture, or designed fraud, may make them almost worthless, with the impossibility of detecting the imposition, without a chemical analysis, till it becomes too late, and the crop is lost.

It is, therefore, safest to rely mainly upon the home manufacture of manure. The extra expense of soiling cattle, saving and applying the liquid manure, and thus bringing the land to a higher state of cultivation, when it will be capable of keeping more stock, and of furnishing more manure, would offer a surer road to success than a constant outlay for concentrated fertilizers.

The various articles used for top-dressing grass lands, and the management of grass and pasture lands, have been treated of in detail in the work already alluded to, on the CULTURE OF GRASSES AND FORAGE PLANTS.

CHAPTER VII.

MILK.

MILK, as the first and natural food of man, has been used from the remotest antiquity of the human race. It is produced by the females of that class of animals known as the *mammalia*, and was designed by nature as the nourishment of their young; but the richest and most abundant secretions in common use are those of the cow, the camel, the mare, and the goat. The use of camel's milk is confined chiefly to Africa and to China, that of mares to Tartary and Siberia, and that of goats to Italy and Spain. The milk of the cow is universally esteemed.

Milk is an opaque fluid, generally white in color, having a sweet and agreeable taste, and is composed of a fatty substance, which forms butter, a caseous substance, which forms cheese, and a watery residuum, known as serum, or whey, in cheese-making. The fatty or butyraceous matter in pure milk varies usually from two and a half to six and a half per cent.; the caseous or cheesy matter, from three to ten per cent.; and the serous matter, or whey, from eighty to ninety per cent.

To the naked eye milk appears to be of the same character and consistence throughout; but under the microscope a myriad of little globules of varied forms, but mostly round or ovoid, and of very unequal sizes, appear to float in the watery matter. On more minute examination, these butter-globules are seen to be enclosed in a thin film of caseous matter. They are so minute that they filter through the finest paper. Milk readily assimilates with water and other sweet and unfermented liquids, though it weighs four per cent. more than water. Cold condenses, heat liquefies it.

The elements of which it is composed, not being similar in character or specific gravity, undergo rapid changes when at rest. The oily particles, being lighter than the rest, soon begin to separate from them, and rise to the surface in the form of a yellowish semi-liquid cream, while the greater specific gravity of the serous matter, or whey, carries it to the bottom.

A high temperature very soon develops acidity, and hastens the separation of the cheesy matter, or curd, from the whey. And so the three principal elements are easily distinguished.

But the oily or butyraceous matter, in rising to the surface, brings up along with it many cheesy particles, which mechanically adhere to it, and give it more or less of a white instead of a yellow color; and many watery or serous particles, which make it thinner, or more liquid, than it otherwise would be. If it rose up free from the adhesion of the other elements, it would appear in the form of pure butter, and would not need to undergo the process of churning to separate it from other substances. The time may come when some means will be devised, either mechanical or chemical, to separate the butter particles from the rest instantaneously and completely, and thus avoid the often long and tedious process of churning.

The coagulation, or collecting together of the cheesy particles, by which the curd becomes separated from the whey sometimes takes place so rapidly, from the

effect of great heat, or sudden changes in the atmos·
phere, that there is not time for the butter particles to
rise to the surface, and they remain mixed up with the
curd.

Nor does the serous or watery matter remain distinct or free from the mixture of particles of the cheesy
and buttery matters. It also holds in suspension some
alkaline salts and sugar of milk, to the extent of from
three to four per cent. of its weight.

We have, then,

$$\text{Milk.}\begin{cases}\text{Cream.}\begin{cases}\text{Butter.}\\\text{Butter-milk.}\end{cases}\text{Water.}\\\text{Skimmed milk.}\begin{cases}\text{Curd.}\\\text{Whey.}\end{cases}\begin{cases}\text{Buttery and cheesy residuum.}\\\text{Sugar of milk.}\\\text{Salts.}\end{cases}\text{Pure water.}\end{cases}$$

It may be stated, in other words, that milk is composed chiefly of caseine, or curd, which gives it its
strength, and from which cheese is made ; a butyraceous or oily substance, which gives it its richness ; a
sugar of milk, to which it owes its sweetness, and a
watery substance, which makes it refreshing as a beverage ; together with traces of alkaline salts, from whence
are derived its flavor and medicinal properties ; and
that these constituents appear in proportions which
vary in different specimens, according to the breed
of the animal, the food, the length of time after parturition, etc.

Milk becomes sour, on standing exposed to a warm
atmosphere, by the change of its sugar of milk into an
acid known as lactic acid ; and it is owing to this sugar,
and the chemical changes to which it gives rise, that milk
is susceptible of undergoing all degrees of fermenta·
tion, and of being made into a fermented and palatable
but intoxicating liquor, which, by distillation, produces
pure alcohol. This liquor is extensively used in some

countries. The arrack of the Arabs is sometimes made from camel's milk.

The Tartars make most of their spirituous liquors from milk; and for this purpose they prefer mare's milk, on account of its larger percentage of sugar, which causes a greater and more active fermentation. The liquor made from it is termed milk-wine, or khoumese. It resembles beer, and has intoxicating qualities. The process of manufacture is very simple. The milk, being allowed first to turn sour, is then heated to the proper temperature, when it begins to ferment; and in a day in summer, or two or three days in winter, the process is completed, and the liquor may be kept several weeks without losing its good qualities.

The admirable though complicated organization of the udder and teats of the cow has already been explained, in speaking of the manner of milking. But it may be said, in general, that the number of stomachs or powerful digestive organs of the ruminants is wonderfully adapted to promote the largest secretions of every kind.

The udder of the cow, the more immediate and important receptacle of milk, and in which other milk-vessels terminate, is divided into two sections, and each of these sections is subdivided into two others, making four divisions, each constituting in itself, to some extent, an organ of secretion. But it is well known that, as a general thing, the lateral section, comprising the two hind teats, usually secretes larger quantities of milk than the front section, and that its development, both external and internal, is usually the greatest.

Milk is exceedingly sensitive to numerous influences, many of which are not well understood. It is probably true that the milk of each of the divisions of the udder differs to some extent from that of the others in the same animal; and it is well known that the milk of different cows, fed on the same food, has marked differences in quality and composition. But food, no doubt, has a more powerful and immediate effect than anything else, as we should naturally suppose from the fact that it goes directly to supply all the secretions of the body. Feeding exclusively on dry food, for instance, produces a thicker, more buttery and cheesy milk, though less abundant in quantity, than feeding on moist and succulent food. The former will be more nutritive than the latter.

Cows in winter will usually give a milk much richer in butter and less cheesy than in summer, for the same reason; while in summer their milk is richer in cheese and less buttery than in winter. As already intimated, the frequency of milking has its effect on the quality. Milking but once a day would give a more condensed and buttery milk than milking twice or three times. The separation of the different constituents of milk begins, undoubtedly, before it leaves the udder; and hence we find that the milk first drawn from the cow at a milking is far more watery than that drawn later, the last drawn, commonly called the strippings, being the richest of all, and containing from six to twelve times as much butter as the first.

Many other influences affect the milk of cows, both in quantity and quality, as the length of time after calving, the age and health of the cow, the season of the year, etc. Milk is whiter in color in winter than in summer, even when the feeding is precisely the same. At certain seasons the milk of the same cow is bluer than at others. This is often observable in dog-days.

The specific gravity of milk is greater than that of water, that of the latter being one thousand, and that of the former one thousand and thirty-one on an average, though it varies greatly as it comes from different cows, and even at different times from the same cow. A feed-ing of salt given to the cow will, in a few hours, cause the specific gravity of her milk to vary from one to three per cent.

Milk will ordinarily produce from ten to fifteen per cent. of its own volume in cream ; or, on an average, not far from twelve and a half per cent. Eight quarts of milk will, therefore, make about one quart of cream. But the milk of cows that are fed so as to produce the richest milk and butter will often very far exceed this, sometimes giving over twenty per cent. of cream, and in very rare instances twenty-five or twenty-six per cent. The product of milk in cream is more regular than the product of cream in butter. A very rich milk is lighter than milk of a poor quality, for the reason that cream is lighter than skim-milk.

Of the different constituents of milk, caseine is that which most resembles animal matter, and hence the intrinsic value of cheese as a nutritive article of food. Hence, also, the nutritive qualities of skimmed milk, or milk from which the cream only has been removed, while the milk is still sweet. The oily or fatty parts of milk furnish heat to the animal system ; but this is easily supplied by other substances.

From the peculiar nature of milk, and its extreme sensitiveness to external influences, the importance of the utmost care in its management must be apparent : and this care must begin from the moment when it leaves the udder, especially if it is to be made into butter. In this case it would be better, if it were con-venient, to keep the different kinds of milk of the same milking by itself — that which comes first from the udder, and that which is drawn last ; and if the first third could be set by itself, and the second and the third parts

by themselves, the time required to raise the cream of
each part would doubtless be considerably less than it
is where the different elements of the milk are so intimately mixed together in the process of milking, after
being once partially separated, as they are before they
leave the udder.

After milking, as little time as possible should elapse
before the milk is brought to rest in the pan. The
remarks of Dr. Anderson on the treatment of milk are
pertinent in this connection. " If milk," says he, " be
put into a dish and allowed to stand until it throws up
cream, the portion of cream rising first to the surface
is richer in quality and equal in quantity to that which
rises in a second equal space of time ; and the cream
which rises in a second interval of time is greater in
quantity and richer in quality than that which rises in a
third equal space of time. That of the third is greater
than that of the fourth, and so of the rest; the cream
that rises continuing progressively to decrease in
quantity and quality, so long as any rises to the surface.

" Thick milk always throws up a much smaller proportion of the cream which it actually contains than
milk that is thinner, but the cream is of a richer quality ; and if water be added to that thick milk, it will
afford a considerably greater quantity of cream, and
consequently more butter, than it would have done if
allowed to remain pure ; but its quality at the same time
is greatly deteriorated.

" Milk which is put into a bucket or other proper
vessel, and carried in it to a considerable distance, so as
to be much agitated and in part cooled before it be put
into the milk-pans to settle for cream, never throws up
so much or so rich a cream as if the same milk had been
put into the milk-pans, without agitation, directly after it
was milked."

18

Milk as it comes from the cow is about blood-heat, or 98° Fah. It should be cooled off as little as possible before coming to rest. ˈ With this object in view, the pails may be rinsed-with hot water before milking, and the distance from the place of milking to the milk-room should be as short as possible ; but, even with all these precautions, the fall in temperature will be considerable.

From what has already been said with regard to the manner in which the cream or oily particles of the milk rise to the surface, and the difficulty of rising through a great space, on account of their intimate entanglement with the cheesy and other matters, the importance of using shallow pans must be sufficiently obvious.

To facilitate and hasten the rising of the butter or oily particles, the importance of keeping the milk-room at a uniform and pretty high temperature will be equally obvious. The greatest density of milk is at or near the temperature of 41° Fah.; and at this point the butter particles will, of course, rise with the greatest difficulty and slowness, and bring up a far greater amount of cheese particles than under more favorable circumstances. These caseous and watery matters, as has been already stated, cause the cream or the butter to look white, and to ferment and become rancid. To avoid this, the temperature is generally kept, in the best butter-dairies, as high as from 58° to 62°. Some recommend keeping the milk at over 70°, and from that to 80°, at which temperature the cream, they say, rises very rapidly, especially if the depth through which it has to rise is but slight. But that, in the opinion of most practical dairymen, is too high.

To obtain the greatest amount of cream from a given quantity of milk, the depth in the pan should, it seems to me, never exceed two inches. A high temperature and shallow depth, as they liquefy the milk and facilitate the rising of the particles, tend to secure a cream free from the cheesy matter, and such cream will make a quality of butter both more delicate to the taste, and less likely to become rancid, than any other.

It has already been intimated, in another connection, that neither the largest quantity nor the best quality of milk is given by the cow till after she has had two or three calves, or has arrived at the age of five or six years. It may also be said, what cannot fail to have attracted the attention of observing dairymen, that in very dry seasons the quantity of milk yielded will generally be less, though the quality will be richer, than in moist and mild seasons.

Hence it may be inferred that moist climates are much more favorable to the production of milk than dry ones ; and this also has been frequently observed and admitted to be a well-known fact. From these facts it may be stated that dry and warm weather increases the quantity of butter, but it is also true that cooler weather produces a greater amount of cheese. A state of pregnancy, it is obvious, must reduce the quality of the milk, and cause it to yield less cream than before.

In the treatment of milk the utmost cleanliness is especially requisite. The pails, the strainers, the pans, the milk-room, and, in short, everything connected with the dairy, must be kept neat and clean to an extent which few but the very best dairy-women can appreciate. The smallest portion of old milk left to sour in the strainers or pans will be sure to taint them, and impart their bad flavor to the new milk put into them. Every one is familiar with the fact that an exceedingly small quantity of yeast causes an active fermentation. The process is a chemical one, and another familiar instance of it is in the distillation of liquors and the brewing of beer, where the malt creates a very active fermentation. In a similar manner the smallest particle of sour milk will taint a large quantity of sweet.

The milk-room should be removed from dampness, and all gases which might be injurious to the milk by infecting the atmosphere. If the state of the atmosphere and the temperature, as has been stated, affect it, all contact with foreign substances to which it is liable in careless and slovenly milking, and all air rendered impure by vegetables and innumerable other things kept in a house-cellar, will be much more liable to taint and injure it. Milk appears to absorb odors from objects near it, to such an extent that a piece of catnip lying near the pan has been known to impart its flavor to it.

Milk, as sold in most large cities, is often adulterated to a great extent, but most frequently with water. Not unfrequently, too, a part of the cream is first taken off, and water afterwards added; in which case the use of burnt sugar is very common for coloring the milk, the blueness of which would otherwise lead to detection. The adulteration of pure milk from the healthy cow by water, though dishonest, and objectionable in the highest degree, is far less iniquitous in its consequences than the nefarious traffic in "swill-milk," or milk produced from cows fed entirely on "still-slops," from which they soon become diseased, after which the milk contains a subtle poison, which is as difficult of detection by any known process of chemistry as the miasma of an atmosphere tainted with yellow fever or the cholera. The simple fact is sufficiently palpable, that no pure and healthy milk can be produced by an unhealthy and diseased animal; and that no animal can long remain healthy that is fed on an unnatural food, and treated in the manner too common around the distilleries of many large cities.

It is evident, from the well-known influence which "still-slops" and other exceedingly succulent food have in increasing the amount of water in the milk, that adul-teration may be effected by means of the food, as well as by addition of water to the milk itself. It is evident, too, on a moment's reflection, that the specific gravity of pure milk must vary exceedingly, as it comes from different cows, or from the same cow at different times. This variation reached to the extent of twenty-three degrees in the milk of forty-two different cows, or from one thousand and eight to one thousand and thirty-one; but so great a variation is very rare, and not to be expected.

No reliable conclusion, as to whether a particular specimen of milk has been adulterated or not, can there-fore be drawn from the differences in specific gravity alone. A radical difficulty attending this test arises from the fact that the specific gravity both of water and cream is less than that of pure milk. If, therefore, the hydrometer sinks deeper into the fluid than would be expected in ordinary pure milk, how is it possible, unless the variation is very large, to tell whether it is due to the richness of the milk in cream, or to the water? I have, for instance, two instruments, each labelled "Lactometer," but both of which are simple hydrometers (Fig. 71), or specific gravity testers, one of which is graduated with the water-mark 0 and that of pure milk 20°; the water-mark of the other being 0, like the first, and that of pure milk 100°. Both are the same in principle, the only difference being in the graduation. On the former, graduated for pure milk at 20°, it is difficult to tell with accuracy the small variations in

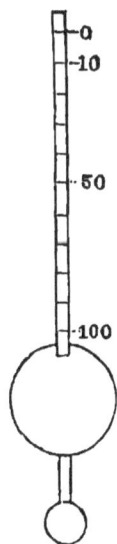
Fig. 71.

18*   14

the percentage of water or cream, the divisions on the scale are so minute, while the latter marks them so that they can be read off with greater ease and precision.

For the purpose of showing the difference in the specific gravity in different specimens of pure milk, taken from the cows in the morning, and allowed to cool down to about 60°, 1 used the latter instrument with the following results: The first pint drawn from a native cow stood at 101°, the scale being graduated at 100° for pure milk. The last pint of the same milking, being the strippings of the same cow, stood at 86°. The mixture of the two pints stood at about 93½°. The milk of a pure-bred Jersey stood at 95°, that of an Ayrshire at 100°, that of a Hereford at 106°, that of a Devon at 111°, while a thin cream stood at 66°. All these specimens of milk were pure, and milked at the same time in the morning, carefully labelled in separate vessels, and set upon the same shelf to cool off; and yet the variations of specific gravity amounted to 25°, or, taking the average quality of the native cows' milk at 93½°, the variations amounted to 17½°.

But, knowing the specific gravity, at the outset, of any specimen of milk, the hydrometer would show the amount of water added. This cheap and simple instrument is therefore of frequent service.

The lactometer is a very different instrument, and measures the comparative richness of different specimens of milk. It is of very great service both in the butter and cheese dairy, for testing the comparative value of different cows for the purposes for which they are kept. This instrument is very simple and cheap, and the practical dairyman can tell by it what cows he can best part with without detriment to his business.

No cow should be admitted to a herd kept for butter-making without knowing her qualities in this respect.

Many would find, on examination, that some of their cows, though giving a good quantity, were comparatively worthless to them. Such was the experience of John Holbert, of Chemung, New York, who, in his statement to the state agricultural society, says: " I find, by churning the milk of each cow separately, that one of my best cows will make as much butter as three of my poorest, *giving the same quantity* of milk. I have kept a dairy for twenty years, but I never until the past season knew that there was so much difference in cows."

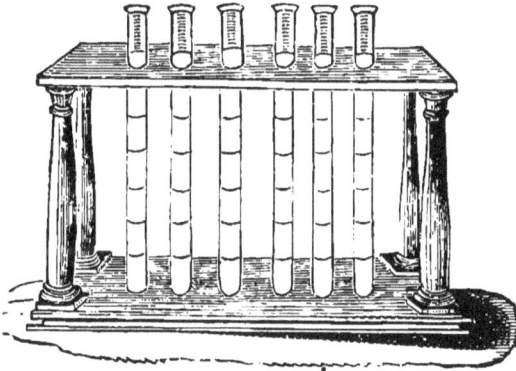

Fig. 72. Lactometer.

The simplest form of the lactometer is a series of graduated glass tubes (Fig. 72), or vials, of equal diameter; generally a third of an inch inside, and about eleven inches long. The tubes are filled to an equal height, each one with the milk of a different cow, and allowed to stand for the cream to rise. The difference in thickness of the column of cream will be very perceptible, and it will be greater than most people imagine. The effect of different kinds of food for the production of butter may be studied in the same way.

This form of the lactometer was invented by Sir Joseph Banks.

Various means are used for the preservation of milk. One of these is by concentrating it by boiling. Where this is followed, as it is by some dairymen, as a regular business, the milk is poured, as it comes from the dairy, into long, shallow, copper pans, and heated to a temperature of a hundred and ten degrees, Fahrenheit. A little sugar is then mixed in, and the whole body of milk is kept in motion by stirring for some three or four hours. The water is evaporated, leaving the milk about one fourth of its original bulk. It is now put into tin cans, the covers of which are soldered on, when the cans are lowered into boiling water. After remaining a while, they are taken out and hermetically sealed, in which condition the milk will keep for months. Concentrated milk may thus be taken to sea or elsewhere. Another form is that of solidified milk, in which state it is easily and perfectly soluble in water; and when so dissolved with a proper proportion of water, it assumes its original form of milk, and may be made into butter. A statement by Dr. Dorémus, in the New York *Medical Journal*, explains the process, as follows:

To one hundred and twelve pounds of milk twenty eight pounds of Stuart's white sugar were added, and a trivial portion of bicarbonate of soda, — a teaspoonful, — merely enough to insure the neutralizing of any acidity, which, in the summer season, is exhibited even a few minutes after milking, although inappreciable to the organs of taste. The sweet milk was poured into evaporating pans of enamelled iron, imbedded in warm water heated by steam. A thermometer was immersed in each of these water-baths, that, by frequent inspection, the temperature might not rise above the point which years of experience have shown advisable. To facilitate the evaporation, by means of blowers and other ingenious apparatus a current of air is established between the covers of the pans and the solidifying milk. Connected with the steam-engine is an arrangement of stirrers, for agitating the milk slightly, while evaporating, and so gently as not to *churn* it. In about three hours the milk and sugar assumed a pasty consistency, and delighted the palates of all present. By constant manipulation and warming, it was reduced to a rich, creamy-looking powder, then exposed to the air to cool, weighed into parcels of a pound each, and by a press, with the force of a ton or two, made to assume the compact form of a tablet (the size of a small brick), in which shape, covered with tin-foil, it is presented to the public.

"Some of the solidified milk which had been grated and dissolved in water the previous evening was found covered with a rich cream; this, skimmed off, was soon converted into excellent butter. Another solution was speedily converted into wine-whey by a treatment precisely similar to that employed in using ordinary milk. It fully equalled the expectations of all; so that solidified milk will hereafter rank among the necessary appendages to the sick room. In fine, this article makes paps, custards, puddings, and cakes, equal to the best milk; and one may be sure it is an unadulterated article, obtained from well-pastured cattle, and not the produce of distillery slops; neither can it be *watered*. For our steamships, our packets, for those travelling by land or by sea, for hotel purposes, or use in private families, for young or old, we recommend it cordially as a substitute for fresh milk."

A pound of this solidified milk, it is said, will make five pints when dissolved in water.

Another favorite form in which milk is used is that

known as ice-cream, a cheap and healthy luxury during the summer months. It is frozen in a simple machine made for the purpose, in the best form of which the time of the operation is from six to ten minutes. The richest quality of ice-cream is made from cream, in the following manner: To one quart of cream use the yolks of three eggs. Put the cream over the fire till it boils, during which time the eggs are beaten up with half a pound of white sugar, powdered fine; and when the cream boils stir it upon the eggs and sugar, then let it stand till quite cold, then add the juice of three or four lemons. It is then ready to put into the freezer. The heat of the cream partially cooks the eggs, and the stirring must be continued to prevent their cooking too much.

A somewhat simpler receipt, given by the confectioners, is the following: To half a pound of powdered sugar add the juice of three lemons. Mix the sugar and lemon together, and then add one quart of cream. This is less rich and delicate than the preceding, but is quite rich enough for common use, and some trouble is saved.

The following receipt makes a very good ice-cream.

Two quarts of good *rich* milk; four fresh eggs; three quarters of a pound of white sugar; six teaspoons of Bermuda arrow-root. Rub the arrow-root smooth in a little cold milk, beat the eggs and sugar together, bring the milk to the boiling point, then stir in the arrow-root; remove it then from the fire, and immediately add the eggs and sugar, stirring briskly, to keep the eggs from cooking, then set aside to cool. If flavored with extracts, let it be done *just before* putting it in the freezer. If the vanilla bean is used, it must be boiled in the milk. The preparation must be *thoroughly cooled* before the freezing is proceeded with.

The ice-cream by this receipt may be produced at a

cost not exceeding twenty-five cents a quart, calling the milk five cents a quart, and the eggs a cent apiece, and including the cost of labor. It is quite equal to that commonly furnished by the confectioners at seventy-five cents a quart. The arrow-root may be dispensed with. The freezer is a cheap and simple machine.

After the cream has frozen in the machine, it should stand an hour or two to harden before it is used.

To secure a more uniform flow and a richer quality of milk, cows are sometimes spayed, or castrated. The milk of spayed cows is pretty uniform in quantity, and this quantity will be, on an average, a little more than before the operation was performed. But few instances have come under my observation, and those few have resulted satisfactorily, the quality of the milk having been greatly improved, the yield becoming regular for some years, and varying only by the difference in the succulence of the food. The proper time for spaying is about five or six weeks after calving, or at the time when the largest quantity of milk is given. There seem to be some advantages in spaying for milk and butter dairies, where the raising of stock is not attended to. The cows are more quiet, never being liable to returns of seasons of heat, which always more or less affect the milk both in quantity and quality. They give milk nearly uniform in these respects, for several years, provided the food is uniformly succulent and nutritious. Their milk is influenced like that of other cows, though to less extent, by the quality and' quantity of food; so that in winter, unless the animal is properly attended to, the yield will decrease somewhat, but will rise again as good feed returns. This uniformity for the milk-dairy is of immense advantage. Besides, the cow, when old, and inclined to dry up, takes on fat with greater rapidity, and produces a juicy and tender beef, superior, at the same age, to that of the ox. The operation of spaying is simple, and may be performed by any veterinary surgeon, without much risk of injury.

The milk of the cow has often been analyzed. It was found by Haidlen to consist of

Water, ...	873.	Magnesia, ...	.42
Butter, ...	30.	Iron, ...	.47
Caseine, ...	48.2	Chloride of Potassium, . .	1.44
Sugar of milk, . . .	43.9	Sodium and Soda, . . .	.66
Phosphate of lime, . .	2.31		1000.

But its composition, as already intimated, varies exceedingly with the food of the animal, and is influenced by an infinite variety of circumstances.

Skim-milk is much more watery than whole milk. It was found by one analysis to contain about 97 per cent. of water and 3 per cent. of caseine.

Swill-milk, or milk from cows fed on "still-slops," in New York, was found by analysis to contain less than 1.5 per cent. of butter, some specimens having even less than one per cent.

The colostrum, or milk of the cow just after calving, contains a large proportion of cheesy matter. Its amount of caseine was found by careful analysis to be 15.1 per cent., of butter 2.6, mucous matter 2, and water 80.3, there being only a trace of sugar of milk.

The measures for milk in common use in this country are those used for wine and beer. The wine quart is about one fifth less than the beer quart, and is that most commonly used in England. It is to be regretted that no uniform standard has been adopted throughout the country.

CHAPTER VIII.

BUTTER AND THE BUTTER-DAIRY.

" Slow rolls the churn — its load of clogging cream
At once foregoes its quality and name.
From knotty particles first floating wide,
Congealing butter 's dashed from side to side."

BUTTER, as we have seen, is the oily or fatty con-
stituent of all good milk, mechanically united or held in
suspension by the solution of caseine or cheesy matter
in water. It is already formed in the udder of the cow,
and the operations required after it leaves the udder, to
produce it, effect merely the separation, more or less
complete, of the butter from the cheese and the whey.

This being the case, it is natural to suppose that
butter was known at an early date. The wandering
tribes, accustomed to take on their journeys a supply
of milk in skins, would find it formed by the agitation
of travelling, and thus would be suggested the first
rude and simple process of churning.

But it is not probable that the Jews possessed a
knowledge of it; and it is pretty well settled, at the
present time, that the passages in our English version
of the Old Testament in which it is used are errone-
ously translated, and that wherever the word butter
occurs the word milk, or sour, thick milk, or cream,
should be substituted. And so in Isaiah, " Milk and
honey shall he eat," instead of " butter; " and in Job
(29 : 6), " When I washed my feet in milk," instead of

19

"butter." And the expression in Prov. (30 : 33), "Surely the churning of milk bringeth forth butter," would be better translated, according to the best critics, "the pressing of the milker bringeth forth milk," or the "pressing of milk bringeth forth cheese."

In the oldest Greek writers milk and cheese are spoken of, but there is no evidence that butter was known to them. The Greeks obtained their knowledge of it from the Scythians or the Thracians, and the Romans obtained theirs from the Germans.

In the time of Christ it was used chiefly as an ointment in the baths, and as a medicine. In warm latitudes, as in the southern part of Europe, even at the present day, its use is comparatively limited, the delicious oil of the olive supplying its place.

I have already stated that all good milk of the cow contained butter enclosed in little round globules held in suspension, or floating in the other substances. As soon as the milk comes to rest after leaving the udder, these round particles, being lighter than the mass of cheesy and watery materials by which they are surrounded, begin to rise and work their way to the surface. The largest globules, being comparatively the lightest, rise first, and form the first layer of cream, which is the best, since it is less filled with caseine. The next smaller, rising a little slower, are more entangled with other substances, and bring more of them to the surface ; and the smallest rise the slowest and the last, and come up loaded with foreign substances, and produce an inferior quality of cream and butter. The most delicate cream, as well as the sweetest and most fragrant butter, is that obtained by a first skimming, only a few hours after the milk is set. Of three skimmings, at six, twelve, and eighteen hours after the milk is strained into the pan, that first obtained will make more and richer butter than the second, and that next obtained richer than the third, and so on.

The last quart of milk drawn at a milking, for reasons already stated, will make a more delicious and savory butter than the first; and if the last quart or two of a milking is set by itself, and the first cream that rises taken from it after standing only five or six hours, it will produce the richest and highest-flavored butter the cow is capable of giving, under like circumstances as to season and feed.

The separation of the butter particles from the others is slower and more difficult in proportion to the thickness and richness of the milk. Hence in winter, on dry feeding, the milk being richer and more buttery, the cream or particles of butter are slower and longer in rising. But, as heat liquefies milk, the difficulty is overcome in part by elevating the temperature. The same effect is produced by mixing a little water into the milk when it is set. It aids the separation, and consequently more cream will rise in the same space of time, from the same amount of rich milk, with a little water in it, than without. Water slightly warm, if in cold weather, will produce the most perceptible effect. The quantity of butter will be greater from milk treated in this way; the quality, slightly deteriorated.

It must be apparent, from what has been said, that butter may be produced by agitating the whole body of the milk, and thus breaking up the filmy coatings of the globules, as well as by letting it stand for the cream to rise. This course is preferred by many practical dairymen, and is the general practice in some of the countries most celebrated for superior butter.

The general treatment of milk and the management of cream have been already alluded to in a former chapter. It has been seen that the first requisites to successful dairy husbandry are good cows, and abundant and good feeding, adapted to the special object of the dairy, whether it be milk, butter, or cheese; and that, with both these conditions, an absolute cleanliness in every process, from the milking of the cow to bringing the butter upon the table, is indispensably necessary.

Cleanliness may, indeed, with propriety be regarded as the chief requisite in the manufacture of good butter; for the least suspicion of a want of it turns the appetite at once, while both milk and cream are so exceedingly sensitive to the slightest taint in the air, in everything with which they come in contact, as to impart the unmistakable evidence of any negligence, in the taste and flavor of the butter.

It is safe to say, therefore, that good butter depends more upon the manufacture than upon any other one thing, and perhaps than all others put together. So important is this point, that a judicious writer remarks that " in every district where good butter is made it is universally attributed to the richness of the pastures, though it is a well-known fact that, take a skilful dairymaid from that district into another, where good butter is not usually made, and where, of course, the pastures are deemed very unfavorable, she will make butter as good as she used to do. And bring one from this last district into the other, and she will find that she cannot make better butter there than she did before, unless she takes lessons from the servants, or others whom she finds there ; " and a French writer very justly observes that " the particular nature of Bretagne butter, whose color, flavor, and consistence, are so much prized, depends neither on the pasture nor on the particular species of cow, but on the mode of making; " and this will hold to a considerable extent, in every country where butter is made.

Many things, indeed, concur to produce the best results, and it would be useless to underrate the importance of any ; but, with the best of cows to impart the
proper color and consistency to butter, the sweetest
feed and the purest water to secure a delicate flavor,
the utmost care must still be bestowed by the dairymaid
upon every process of manufacture, or else the best of
milk and cream will be spoiled, or produce an article
which will bring only a low price in the market, when,
with greater skill, it might have obtained the highest.

From what has been said of the care requisite to preserve the milk from taint, it may be inferred that attention to the milk and dairy room is of no small importance.
In very large butter-dairies, a building is devoted exclusively to this department. This should be at a short
distance from the yard, or place of milking, but no
further than is necessary to be removed from all impurities in the air arising from it, and from all low, damp
places, subject to disagreeable exhalations. This is of
the utmost importance. It should be well ventilated,
and kept constantly clean and sweet, by the use of pure
water; and especially, if milk is spilled, it should be
washed up immediately, with fresh water. No matter if
it is but a single drop; if allowed to soak into the floor
and sour, it cannot easily be removed, and it is sufficient
to taint the air and the milk in the room, though it may
not be perceptible to the senses.

In smaller dairies, economy dictates the use of a room
in the house ; and this, in warm climates, should be on
the north side, and used exclusively for this purpose.
I have known many to use a room in the cellar as a
milk-room ; but very few cellars are at all suitable.
Most are filled with a great variety of articles which
never fail to infect the air.

But, if a house-cellar is so built as to make it a suita-
19*

ble place to set the milk, as where a large dry and airy room, sufficiently isolated from the rest, can be used, a greater uniformity of temperature can usually be secured than on the floor above. The room, in this case, should have a gravel or loamy bottom, uncemented, but dry and porous. The soil is a powerful absorbent of the noxious gases which are apt to infect the atmosphere near the bottom of the cellar.

Milk should never be set on the bottom of a cellar, if the object is to raise the cream. The cream will rise in time, but rarely or never so quickly or so completely as on shelves from five to eight feet from the bottom, around which a free circulation of pure air can be had from the latticed windows. It is, perhaps, safe to say that as great an amount of better cream will rise from the same milk in twelve hours on suitable shelves, six

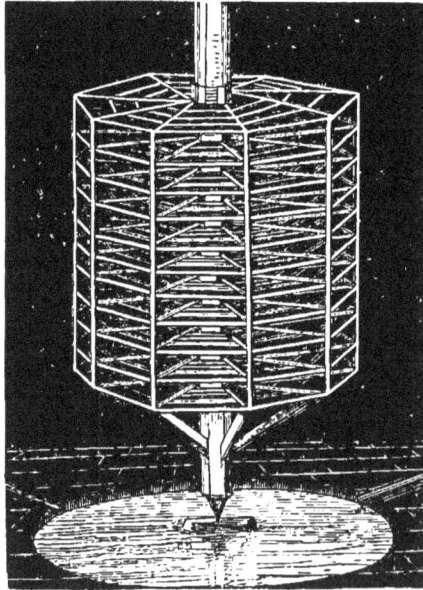

Fig. 73. Milk-stand.

feet from the bottom, as would be obtained directly on the bottom of the same cellar in twenty-four hours.

One of the most convenient forms for shelves in a dairy-room designed for butter-making is represented in Fig. 73, made of light and seasoned wood, in an octagonal form, and capable of holding one hundred and seventy-six pans of the ordinary form and size. It is so simple and easily constructed, and so economizes space, that it may readily be adapted to other and smaller rooms for a similar purpose. If the dairy-house is near a spring of pure and running water, a small stream can be led in by one channel and taken out by another, and thus keep a constant circulation under the milk-stand, which may be so constructed as to turn easily on the central post, so as often to save many footsteps.

The pans designed for milk are generally made of tin. That is found, after long experience, to be, on the whole, the best and most economical, and subject to fewer objections than most other materials. Glazed earthen ware is often used, the chief objection to it being its liability to break, and its weight. It is easily kept clean, however, and is next in value to tin, if not, indeed, equal to it. A tin skimmer is commonly used, somewhat in the form of the bowl of a spoon, and pierced with holes, to remove the cream. In some sections of the country, a large white clam-shell is very commonly used instead of a skimmer made for the purpose, the chief objection to it being that the cream is not quite so carefully separated from the milk.

A mode of avoiding the necessity of skimming has long been used to some extent in England, by which the milk is drawn off through a hole in the bottom of the pan. This plan is recommended by Unwerth, a German agriculturist, who proposes a pan represented in Fig. 74, made of block tin, oblong in shape, and having the inside corners carefully rounded. The pan is only two inches in depth, and is made large enough to hold six or eight quarts of milk at the depth of one and a half inches. This shallowness greatly facilitates

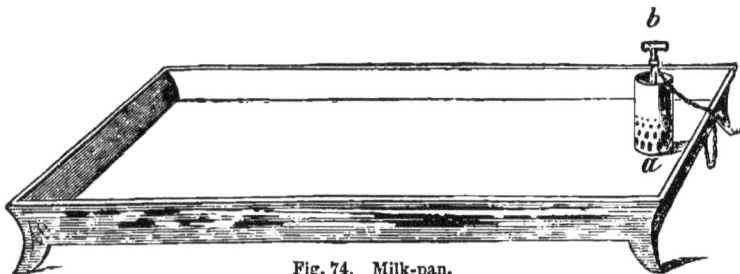

Fig. 74. Milk-pan.

the rapid separation of the cream, especially at a temperature somewhat elevated. A strainer is shown in Fig. 75, pierced with holes, the centre half an inch lower than the rim, to which hooks are fixed to hold it to the top of the pan. On this a coarse linen cloth

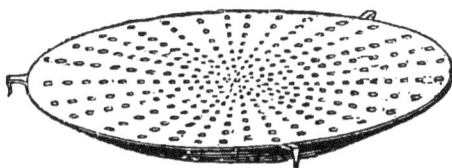

Fig. 75.

Fig. 76.

is laid, the milk being strained through both the cloth and the strainer, thus serving to separate all foreign substances in a thorough manner.

In the bottom of the milk-pan, near one end, is an opening, a, through which the milk is drawn, after the cream is all risen or separated from it, by raising a brass pin, b. The opening is lined with brass, and is three fourths of an inch in diameter. Fig. 76 represents the tin cylinder magnified. This is pierced, to the height of an inch, with many small holes, diminishing in size towards the top. The cream is all risen in twenty four hours. The pin is then drawn from the cylinder, and the milk flows out, leaving the thick cream, which is prevented from flowing out by the smallness of the holes in the cylinder.

With the form of pans in most common use in this country, which are circular, three or four inches deep, this shallow depth of milk causes a little more trouble in skimming; but, if the principle is correct, the form and depth of the pan will be easily adapted to it.

After the cream is removed, it is put into stone or earthen jars, and kept in a cool place till a sufficient quantity is accumulated to make it convenient to churn. If a sufficient number of cows is kept, it is far better to churn every day; but in ordinary circumstances that may be oftener than is practicable. The more frequently the better; and the advantages of frequent churning are so great that cream should never be kept longer than three or four days, where it is possible to churn so often.

The mode of churning in one of the many good dairies in Pennsylvania, — that of Mr. J. Comfort, of Montgomery county, — is as follows: He uses a large barrel-shaped churn, of the size of about two hogsheads, hung on journals supported by a framework in an adjoining building. It is worked by machinery in a rotatory motion, by a horse travelling around in a circle. The churning commences about four o'clock in the morning in summer, the cream being poured into the churn and the horse started. When the butter has come, a part of the butter-milk is removed by a vent-hole in the churn. Then, without beating the mass together, as is usual, a portion of the butter and its butter-milk is taken out by the spatula and placed in the bottom of a tub covered with fine salt, and spread out equally to a proper depth; then the surface of this butter is cov-

15

.ered with salt, and another portion of butter and butter-milk taken from the churn and spread over the salted surface in the same manner, and salted as before, thus making a succession of layers, till the tub is full. The whole is then covered with a white cloth, and allowed to stand a while. A part of this butter, say eight or ten pounds, is then taken from the tub and laid on a marble table (Fig. 80), grooved around the edges, and slightly inclined, with a place in the groove for the butter-milk and whey to escape. It is then worked by a butter-worker or brake, turning on a swivel-joint, which perfectly and completely removes the butter-milk, and flattens out the butter into a thin mass; then the surface is wiped by a cloth laid over it, and the working and wiping repeated till the cloth adheres to the butter, which indicates that the butter is dry enough, when it is separated into pound lumps, weighed and stamped, ready for market. The rest of the butter in the tub is treated in the same way.

It will be seen that this method avoids the ordinary washing with water, not a drop of water being used, from beginning to the end. It avoids also the working by hand, which in warm weather has a tendency to soften the butter. In the space of about an hour a hundred pounds are thus made, and its beautiful color and fragrance preserved. If it happens to come from the churn soft, it hardens by standing a little longer in the brine.

The most common form of the churn in small dairies is the upright or dash-churn, Fig. 77; but many other forms

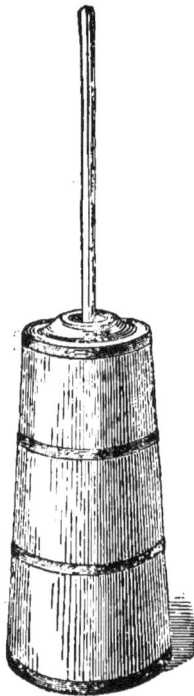

Fig. 77.

are in extensive use, each possessing, doubtless, more
or less merit peculiar to itself. The cylinder churn,

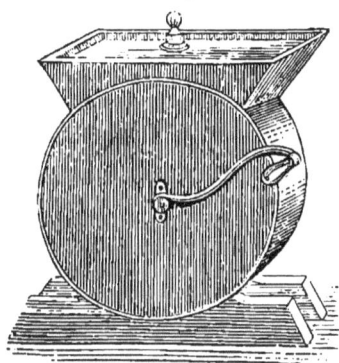

Fig. 78.

Fig. 78, is very simply
constructed, and capable of
being easily cleaned. Some
prefer the thermometer
churn, with a convenient
attachment for indicating
the temperature of the
cream.

As already stated, there
are two modes of practice
with regard to the pro-
cess of churning, each of which has its advantages.
The milk itself may be churned, or it may be set in
the milk-room for the cream to rise.

But, whichever course it is thought best to adopt,
whether the milk or cream is churned, it is the concus-
sion, rather than the motion, which serves to bring the
butter. This may be produced in the simple square box
as well as by the dasher churn ; and it is the opinion of
a scientific gentleman with whom I have conversed on
the subject, that the perfect square is the best form of
the churn ever invented. The cream or milk in this
churn has a peculiar compound motion, and the concus-
sion on the corners and right-angled sides is very great,
and causes the butter to come as rapidly as it is judicious to have it. This churn consists of a simple square
box, which any one who can handle a saw and plane can
make, hung on axles turned by a crank somewhat like
the barrel churn. No dasher is required. If any one
is inclined to doubt the superiority of this form over all
others, he can easily try it and satisfy himself. It costs
but little.

Fig. 79.

This churn, shown in Fig. 79, may be made of any desirable size, but for medium-sized dairies 24 inches each
way, inside, is large enough to churn from 40 to 60 lbs.
of butter in.   The cream is put in through a circular
opening, about seven inches in diameter, having a cover
made to fit tight, and clamped on with screws.   A small
hole is made in one corner at the end, to ventilate the
churn occasionally when in use, stopped by a plug, and
through which the buttermilk is run off when the butter
"has come."   The end of the churn and the motion of
the cream are shown in Fig. 79 a, and the form of the
butter after the buttermilk is worked out in Fig. 79 b.

Mr. N. B. Chamberlain, of Westboro', Mass., who has
this churn in constant use, says : "We churned once a
week *during the winter*, the butter varying from fifteen to
thirty-five pounds, the time from seven to fourteen minutes.   The drawing out of the buttermilk and the work
ing in the salt averaged only five minutes."   The churn

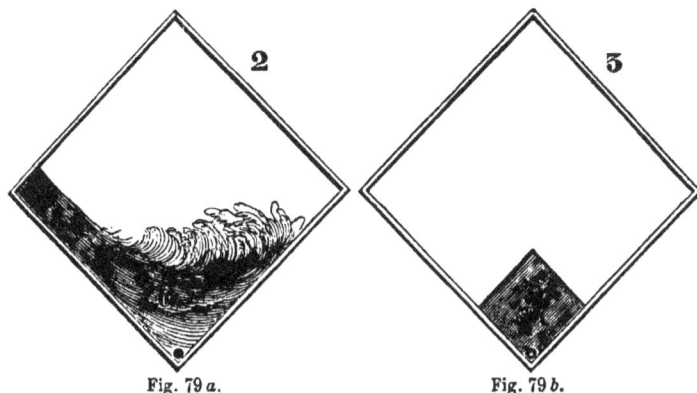

Fig. 79 a.   Fig. 79 b.

ing usually began at 7½ o'clock, and the butter was, with one exception, sent to market before 10 o'clock.

It will be seen that two forces combine to produce the desired result. As the churn revolves, the cream receives all the agitation that is ever given to it by the barrel or cylindrical churn. But, in addition, the cream is dashed violently against the sides of the churn. This dashing motion is the secret of the prompt results secured by the use of the square churn. Moreover, the buttermilk can be drawn out, and the butter not only "gathered," but consolidated, and the buttermilk thoroughly beaten out. If salt is added after drawing out the buttermilk, it is thoroughly worked in in five minutes by giving the churn a half motion.

When the working of this churn is well understood, it will be seen that another great advantage is in churning all the cream. It cannot adhere to the sides and cause streaky butter.

Mr. William S. Lincoln, of Worcester, who received the first premium from the Massachusetts Society for Promoting Agriculture, speaks as follows in regard to the use of the butter-worker: —

"In 'working' butter we use a table over which a fluted roller is made to pass (Fig. 80), rolling out the

20

butter into a thin sheet, and completely and entirely de-
priving it of buttermilk.

"From many years' experience, the observation is
warranted, that by no other process of manufacture
can the butter-milk be so completely extracted. I am
aware of the truth of the objection made that the
shrinkage occasioned by its use is too great; yet there
is, in fact, a difference in the worth of the butter made
upon it, over that manufactured in the ordinary way,
quite equal to the loss in weight occasioned by it."

The high reputation of Philadelphia butter being so
well known, I was desirous of ascertaining the opinions
of practical men as to what this was due,—whether to
any peculiar richness of the pasturage, or to the careful
mode of manufacture.   In reply to my inquiries, I have
received satisfactory statements from several sources,
and among them the following communication from one
of the most successful of the butter-makers who supply
that market.   "The high reputation of Philadelphia butter," he says, "is owing to the manner of its manufacture,
though I would not say that the sweet-scented vernal
and other natural grasses do not add to the fine quality
of well-made butter.

"In proof of what I say, I would refer to the experience of my brother, who is the owner of two farms.
His tenant, an excellent butter-maker, lived on one
farm, and made a very fine article, which brought the
highest prices. He moved to the other farm, where
the former tenant had never made good butter, and had
ascribed his want of success to the spring-house. On
this farm he succeeded in establishing a higher reputation than he ever had before. The tenant who followed him on the first farm never succeeded in gaining
a reputation for good butter, his inability arising from
his ignorance of the proper mode of manufacture, and

his unwillingness to improve by the experience of others.

"Only a part of the information as to the best mode of manufacture can be given, so much depends on the *judgment* and experience of the operator. The first thing required is to provide a suitable place. This should be, for the summer months, a well-ventilated house, over a good spring of water. The second requisite will be proper vessels to hold the milk and cream, and for churning. A table is needed which shall not be used for any other purpose than for working and printing the butter on. I have always used a lever in connection with the table (Fig. 80). A large sponge, with a linen cloth to cover it, with which the milk can be removed from the butter, is another important article; and then a skimmer, either of wood or tin, or both, as may be necessary in the different states of the milk; a thermometer, and a boiler convenient for heating water for cleansing the vessels. No person can expect to make good butter without the greatest attention to the cleanliness of the vessels used for the milk and cream, and care in exposing them to the sun and air.

"After the milk has been brought from the yard or stable, strain it immediately into the pans, in which has been put a little sour milk from which the cream has been removed, the quantity varying from a tablespoonful to half a common teacupful, according to the state of the weather. In very warm weather the smaller quantity is sufficient. But the rule for warm weather will not always hold good; for, from the electrical state of the atmosphere, the milk may sour either too slow or too fast.

"The pans containing the milk should then be set into the water, if the weather be hot: and here is a point where the operator should exercise his or her judgment; for

even in warm weather it may be necessary to draw off
the water from the milk, if the spring be cold. The milk
should remain there, under no circumstances, longer
than the fourth meal, or forty-eight hours; but thirty-
six hours is much to be preferred, if the milk has
become thick, or the cream sufficiently raised, when it
should be taken off carefully, so as not to take any
sour milk with it, and put in the cream-pot. When the
cream-pot is full, sprinkle a small handful of fine salt
over the top of the cream, and let it remain. Our
custom has been, when making butter but once a week,
to pour the cream into a clean vessel at the end of
three days, keeping back any milk that might have been
taken up with the cream, which is found at the bottom
of the jar.

"I would mention that it is essential, in making a fine
article, to keep the cream clear of milk. The next operation will be preparatory to churning, by straining the
cream, and reducing the temperature of the churn by
the use of the cold spring-water. The operation of
churning should neither be protracted nor hastened too
much. After the butter has made its appearance of the
size of a small pea, draw off the milk, and throw in a
small amount of cold water, and gather it. After the
butter has been taken from the churn, it is placed upon
the table, worked over by the lever, and salted; then
worked again with the lever, in connection with the
sponge and cloth, a pan of cold water being at hand,
with a piece of ice in it in summer, into which you
throw the cloth and sponge frequently, and wring out
dry before again using it. These, as well as every
other article which will come into contact with the
butter, must be scalded, and afterward, as well as the
hands, placed in cold water. I would here add that the
use of the sponge is one of the important points in mak

ing butter to keep well; for by it you can remove almost every particle of butter-milk, which is the great agent in the destruction of its sweetness and solidity. For the winter dairy a room in which is placed a stove should be provided, which can be made warm, and also well ventilated. I prefer the use of coal, on account of keeping the fire through the night. My dairy-room is adjoining the spring-house, and connects with it, which I consider important. This room should be used for no other purpose, as cream and butter are the greatest absorbents of effluvia with which I am acquainted. I have known good butter to be spoiled by being placed over night in a close closet.

"The thermometer should always accompany the winter dairy. There is one thing very important in the winter dairy, which, perhaps, I should have placed first, and that is the food of the cows; for, without something else than hay, you will not make very fine butter. Mill-feed and corn-meal I consider about the best for yield and quality, although there are many other articles of food which will be useful, and contribute to the appetite and health of the cattle.

"The process for the winter dairy is similar to that of the summer, with the exception of the regulation as to the temperature of room, etc., which is as follows:

"Particular care should be taken not to let the milk get cold before placing it in the dairy-room; for, should it be completely chilled, the cream will not rise well. Add about a gill of warm water to the sour milk for each pan, before straining into it, which will greatly facilitate the rising of the cream. Keep the temperature of the room as near fifty-eight degrees, Fahrenheit, as possible, and guard against the air being dry by having a small vessel of water upon the stove, or else a dry coat will form on the surface of the cream. The

20*

cream should be kept in a colder place than the dairy-room until the night before churning, when it might be placed in the warm room, so that its temperature shall be about 58°.

"The churn may be prepared by scalding it, and then reduced to the same temperature as the cream by cold water, using the thermometer as a test.

"This regulation of temperature is of the greatest importance: for, should it be too low, you will be a long time churning, and have poor, tasteless butter; if too high, the butter will be soft and white."

What is especially noticeable in the above statement is the use of the sponge, and the thorough and complete removal of all the butter-milk. Here is probably the secret of success, after all. I have given the statement in full, notwithstanding its length, on account of the well-known excellence of the butter produced by the process, as well as for the suggestions with regard to the dairy-rooms, and not because I can recommend all its details for the imitation of others. The use of sour milk in the pans is based, I suppose, on the idea that the cream does not begin to rise till acidity commences in the milk,—an idea which was once pretty generally entertained; but the process of souring undoubtedly commences, though imperceptible to the senses, very soon after the milk comes to rest in the pan. At any rate, there is no doubt that the separation of the butter from the other substances commences at once, and without the addition of any foreign substance to the milk.

Nor do I believe there is any necessity for the milk to stand over twenty-four hours in any case; for I have no doubt that all the best of the cream rises within the first twelve hours, under favorable circumstances, and I am inclined to think that whatever is added to the

quantity of cream after twenty-four hours, detracts
from the quality of the butter to an extent which more
than counterbalances the whole of the quantity.

Many good dairy-women make an exceedingly fine
article, in spite of the defects of some parts of the process of manufacture. This does not show that they
would not make still better butter if they remedied
these defects.

The more we can retard the development of acidity
in the milk, within certain limits, the more cream may
we expect to get; and hence some use artificial means
for this purpose, mixing in the milk a little crystallized
soda, dissolved in twice its volume of water, which
corrects the acidity as soon as it forms. It is a perfectly
harmless addition, and increases the product of the
butter, and improves its quality. But under ordinarily
favorable circumstances, from twelve to eighteen hours
will be sufficient to raise all the cream in summer, and
from twenty to thirty hours in winter.

Fig. 80. Butter-worker.

The butter-worker, Fig. 80, with its marble top,
used by the writer of the statement above, is an im-

portant addition to the implements of the dairy. It

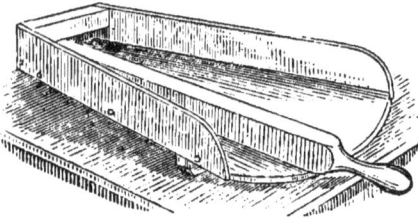

Fig. 81.

effects the complete removal of the butter-milk, without the necessity of bringing the hands in contact with it. Another form of the lever butter-worker is seen in Fig. 81.

To keep the cream properly after it is placed away in pots or jars, it should stand in a cool place, and whenever additions of fresh cream are made, they should be stirred in. Many keep the cream, as well as the butter, in the well, in hot weather. This is the practice of Mr. Horsfall, whose experiments have been alluded to. Finding his butter inclined to be soft, he lowered a thermometer twenty-eight feet into the well, and found it indicated 43°, the temperature of the surface being 70°. He then let down the butter, and found it somewhat improved; and soon after began to lower down the cream, by means of a movable windlass and a rope, the cream-jar being placed in a basket hung on the rope. The cream was let down on the evening previous to churning, and drawn up in the morning and immediately churned. The time of churning the cream at this temperature would be as long as in winter, and the butter was found to have the same consistency.

The same object is effected in this country by the use of ice in many sections ; but, if the butter remains too long on ice, or in an ice-house, it is apt to become bleached, and lose its natural and delicate straw color.

The time of churning is by no means an unimportant matter. Various contrivances have been made to short-en this operation ; but the opinions of the best and most successful dairymen concur that it cannot be too much hastened without injuring the fine quality and consistency of the butter. The time required depends much on the temperature of the cream; and this can be regulated at convenience, as indicated above.

The temperature of the dairy-room should be as uniform as possible. The practice of the best and most successful dairymen differs in respect to the degree to which it should be kept; but the range is from 52° to 62° Fahr., and I am inclined to think from 58° to 60° the best. At 60°, with a current of fresh, pure air passing over it, the cream will rise very rapidly and abundantly.

The greatest density of milk is at about 41°, and cream rises with great difficulty and slowness as the temperature falls below 50° towards that point.

A practical butter dealer of New York gives the following as the best mode of packing butter, or putting it up for a distant market. The greatest care, he says, should be taken to free the butter entirely from milk, by working it and washing it after churning at a temperature so low as to prevent it from losing its granular character and becoming greasy. The character of the product depends in a great measure on the temperature of churning and working, which should be between sixty and seventy degrees Fahr. If free from milk, eight ounces of Ashton salt is sufficient for ten pounds. Western salt should never be used, as it injures the flavor. While packing, the contents of the firkin should be kept from the air by being covered with saturated brine. No undissolved salt should be put in the bottom of the firkin.

Goshen butter is reputed best, though much is put up in imitation of it, and sold at the same price. Great care should be taken to have the firkins neat and clean. They should be of white oak, with hickory hoops, and should hold about eighty pounds. Wood excludes air better than stone, and consequently keeps butter better. Tubs are better than pots.

Western butter comes in coarse, ugly packages; even flour and pork barrels are sometimes used. Much of it must be worked over and re-packed here before it will sell. It generally contains a good deal of milk, and if not re-worked soon becomes rancid. Improper packing, in kegs too large and soiled on the outside, makes at least three cents a pound difference. Whatever the size of the firkin, it must be perfectly tight and quite full of butter, so that when opened the brine, though present, will not be found on the top.

Until the middle of May, dairymen should pack in quarter firkins or tubs, with white oak covers, and send directly to market as fresh butter. From this time until the fall frost there is but little change in color and flavor with the same dairy, and it may be packed in whole firkins, and kept in a cool place. The fall butter should also be packed separately in tubs.

To prepare new butter-boxes for use in the shortest time, dissolve common, or bicarbonate of soda in boiling water, as much as the water will dissolve, and water enough to fill the boxes; about a pound of soda will be required to be put into a thirty-two pound box, and the water should be poured upon it. Let it stand over night, and the box may be safely used next day. This mode is cheap and expeditious, and, if adopted, would often save great losses. Potash has a like effect.

As already seen, in the statements of practical dairy men, the greatest care is required in the salting or seasoning. Over-salted butter is not only less palatable to the taste, but less healthy than fresh, sweet butter. The same degree of care is needed with respect to the box in which it is packed. I have often seen the best and richest-flavored butter spoilt by sending it to the exhibi

tion or to market in new and improper boxes. A new
pine-wood box should always be avoided.

Butter that has been thoroughly worked, and perfectly freed from butter-milk, is of a firm and waxy consistence, so as scarcely to dim the polish of the blade
of a knife thrust into it, leaving upon it only a slight
dew as it is withdrawn. If it is soft in texture, and
leaves greasy streaks of butter-milk upon the knife that
cuts it, or upon the cut surface after the blade is withdrawn, it shows an imperfect and defective process of
manufacture, and is of poor quality, and will be liable
to become rancid.

An exceedingly delicate and fine-flavored butter may
be made by wrapping the cream in a napkin or clean
cloth, and burying it, a foot deep or more, in the earth,
from twelve to twenty hours. This experiment I have
repeatedly tried with complete success, and have never
tasted butter superior to that produced by this method.
It requires to be salted to the taste as much as butter
made by any other process. A tenacious subsoil loam
would seem to be best. After putting the cream into
a clean cloth, the whole should be surrounded by a
coarse towel. The butter thus produced is white
instead of yellow or straw-color.

Butter has been analyzed by Prof. Way, with the following result:

Pure fat, or oil, ... 82.70
Caseine, or curd, ...   2.45
Water, with a little salt, ... 14.85 = 100

The fat or oil peculiar to butter is in winter more
solid than in summer, and known as margarine fat,
while that of summer is known as liquid or oleine fat.
The proportions in which these are found in ordinary

butter have been stated by Prof. Thomson, as follows:

	Summer.	Winter.
Solid or margarine fat, ...	40′	65
Liquid or oleine fat, ...	60	35
	100	100

Winter butter appears to be rich and fine in proportion as the oleine fat increases. The proportion is
undoubtedly dependent on the food.

A more general attention to the details of butter-
making, and to the best modes of preserving its good
qualities, would add many thousands of dollars to the
aggregate profits of our American dairies.

In the management of the dairy, an ice-house and a
good quantity of ice for summer use are not only very
convenient for regulating the temperature of the dairy-
room, and for keeping butter at the proper consistence,
and preserving it, but are also profitable in other respects.
And now, when ice-houses are so easily constructed, and
ice is so readily procured, no well-ordered dairy should
be without a liberal supply of it.   It is housed at a time
when other farm-work is not pressing, and ponds are so
distributed over the country that it may be generally procured without difficulty; but where ponds or streams are
at too great a distance from the dairy-house, an artificial
pond can be easily made, by damming up the outlet of
some spring in the neighborhood.   Where this is done,
the utmost care should be taken to keep the water perfectly clean when the ice is forming.   The ice-house
should be above ground, and in a dry, airy place.   The
top of a dry knoll is better than a low, damp shade.
The ice may be packed in tan, sawdust, shavings, or
other non-conductors, and when wanted for use it
should be taken off the top.

CHAPTER IX.

"Streams of new milk through flowing coolers stray,
And snow-white curds abound, and wholesome whey."

MILK, if allowed to become sour, will eventually curdle, when the whey is easily separated; and this simple mode was probably the universal method of making cheese in ancient times. Cheese, as already explained, is made from caseine, an ingredient of milk held in solution by means of an alkali, which it requires the presence of an acid to neutralize. This, in modern manufacture, is artificially added to form the curd; but the acidity of milk, after standing, acts in the same manner to produce coägulation. This is due to the change of the milk-sugar into lactic acid.

Cheese has been made and used as an article of food from a very early date. It was well known to the early Jewish patriarchs, and is frequently mentioned in the earliest Hebrew records. "Hast thou not poured me out as milk, and curdled me like cheese?" says Job; and David was sent to "carry ten cheeses to the captain of their thousand in the camp." Most of the ancient nations, indeed, barbarous as well as civilized, made it a prominent article of food. But cheese, as made by the ancients, was found to be hard and brittle, and not well flavored, and means were devised to produce the same effect while the milk still remained sweet. It was

21   16

observed that acids of various kinds would answer, and
vinegar was used; and cream of tartar, muriatic acid, and
sour milk, added to sweet, produced a rapid coägulation.
In Sweden, Norway, and other countries, a handful of the
plant known as butterwort (*Pinguicula vulgaris*) is sometimes mixed with the food of the cow, to cause the milk
to coägulate readily. A few hours after milking, the curd
is formed without the addition of an acid. Milk taken
into the stomach of the calf was found to curdle rapidly,
even while sweet; and hence the use of rennet, which is
simply the stomach of the calf, prepared by washing,
salting, and drying, for preservation. This acts the
most surely, and, if properly prepared and preserved, is
the least objectionable, of any article now known; and
is, in fact, the natural mode of curdling the milk as it
enters the stomach, preparatory to the process of digestion. Besides this, it is generally the cheapest and
most available for the farmer.

The richness of cheese depends very much upon the
amount of butter or oily matter it contains. It may be
made entirely of cream, or from whole or unskimmed
milk, to which the cream of other milk is added, or
from milk from which a part of its cream has been
taken, or from ordinary skim-milk, or from milk that
has been skimmed three or four times, so as to remove
nearly every particle of cream, or from butter-milk.
The acid used in curdling milk acts upon the caseine
alone, and not upon the butter particles, which are
imbedded in the curd as it hardens, and thus increase
its richness and flavor without adding to its consistency, which is due to the caseine.

It is evident, therefore, that cheese made entirely of
cream cannot have the firmness and consistence of
ordinary cheese. It is only made for immediate use,
and cannot be long kept. It is, in fact, little more than

thick, dried, sweet cream, from which all the milk has been pressed. On the other hand, skim-milk cheese has the opposite fault of being too hard and tough, and destitute of flavor and richness. The best quality of cheese is made from full milk, or from milk to which some extra cream is added, as in the English Stilton, renowned for its richness and flavor. The Gloucester, Cheshire, Cheddar, Dunlop, and the Dutch Gouda, are made of whole milk, as are the best qualities made in this country.

The process of making cheese is both chemical and mechanical. The heating of the milk at the time of adding the acid or rennet hastens the chemical action, and facilitates the separation of the whey; at the same time great nicety is required, for, if over-heated, the oily particles will run off with the whey. On the complete separation of the whey from the curd, and the amount of butter particles retained in the latter, the taste or flavor and keeping qualities of the cheese depend. If properly made, the taste improves by keeping, but the chemical changes effected by age are not very well understood.

The practical process of manufacture most common in the best dairies of this country will appear in the following statements of successful competitors at agricultural exhibitions. The first was made, by request, to the New York State Agricultural Society, and appeared in its transactions, by A. L. Fish, of Herkimer county, one of the finest dairy regions of that state. The value of his statement is enhanced by the fact that his cows averaged seven hundred pounds of the first quality of cheese each in 1844, and seven hundred and seventy-five pounds each in 1845. In his mode of manufacture, " the evening's and morning's milk is commonly used to make one cheese. The evening's is strained into a tub or pans, and cooled to prevent souring. The proper mode of cooling is to strain the milk into the tin tub set in a wooden vat, described in the dairy-house, and cool by filling the wooden vat with ice-water from the ice-house, or ice in small lumps, and water from the pump. The little cream that rises over night is taken off in the morning, and kept till the morning and evening milk are put together, and the cream is warmed to receive the rennet. It is mixed with about twice its quantity of new milk, and warm water added to raise its temperature to ninety-eight degrees: stir it till perfectly limpid, put in rennet enough to curdle the milk in forty minutes, and mix it with the mass of milk by thorough stirring; the milk having been previously raised to eighty-eight or ninety degrees, by passing steam from the steam generator to the water in the wooden vat. In case no double vat is to be had, the milk may be safely heated to the right temperature, by setting a tin pail of hot water into the milk in the tubs. It may be cooled in like manner by filling the pail with ice-water, or cold spring-water where ice is not to be had. It is not safe to heat milk in a kettle exposed directly to the fire, as a· *slight* scorching will communicate its *taint* to the whole cheese and spoil it. If milk is curdled below eighty-four degrees, the cream is more liable to work off with the whey. An extreme of heat will have a like effect.

The curdling heat is varied with the temperature of the air, or the liability of the milk to cool after adding rennet. The thermometer is the only safe guide in determining the temperature; for, if the dairyman depends upon the sensation of the hand, a great liability to error will render the operation uncertain. If, for instance, the hands have previously been immersed in cold water, the milk will feel warmer than it really is;

if, on the contrary, they. have recently been in warm water, the milk will feel colder than it really is. To satisfy the reader how much this circumstance alone will affect the sensation of the hand, let him immerse one hand in warm water, and at the same time keep the other in a vessel of cold water, for a few moments; then pour the water in the two dishes together, and immerse both hands in the mixture. The hand that was previously in the warm water will feel *cold,* and the other quite warm, showing that the sense of feeling is not a test of temperature worthy of being relied upon. A fine cloth spread over the tub while the milk is curdling will prevent the surface from being cooled by circulation of air. *No jarring of the milk,* by walking upon a springy floor, or otherwise, should be allowed while it is curdling, as it will prevent a *perfect cohesion* of the particles.

"When milk is curdled so as to appear like a solid, it is divided into small particles to aid the separation of the whey from the curd. This is often *too speedily done* to facilitate the work, but at a sacrifice of *quality* and *quantity.*"

To effect the fine division of the curd for the easy separation of the whey, Mr. Fish uses a wire network, made to fit into the tub, the meshes of fine wire being about a half-inch square, and the outer rim of coarse and stronger material. A cheese-knife is also used, about half as long as the diameter of the tub, and firmly fastened to the lower end of a long screw which passes through one end of the blade as it lies horizontally, leaving the blade at right angles with the screw, which has a coarse thread, and passes through a piece of wood on the top of the tub, held firm by notches at the ends laid on the edges of the tub. By turning a crank, the knife passes down through the curd in revolutions

21*

cutting it into layers of the thickness of the threads of the screw.

The following is the statement of Mrs. Williams, of Windsor, Massachusetts, who received the first premium at the Franklin County Fair, in 1857, for exceedingly rich, fine, and delicately-flavored cheeses of seventy-five pounds each. Her method, which is the result of her own experience and observation, corresponds almost exactly, as the committee remark, with the English mode of making the famous Cheddar cheese, which is much the same as the Cheshire. Mrs. Williams says : " My cheese is made from one day's milk of twenty-nine cows. I strain the night's milk into a tub, skim it in the morning, and melt the cream in the morning's milk : I warm the night's milk, so that with the morning's milk, when mixed together, it will be at the temperature of ninety-six degrees ; then add rennet sufficient to turn it in thirty minutes. Let it stand about half or three quarters of an hour; then cross it off and let it stand about thirty minutes, working upon it very carefully with a skimmer. When the curd begins to settle, dip off the whey, and heat it up and pour it on again at the temperature of one hundred and two degrees. After draining off and cutting up, add a teacup of salt to fourteen pounds.

"The process of making sage cheese is the same as the other, except adding the juice of the sage in a small quantity of milk."

Another successful competitor in the same state says : " We usually make but one curd in a day. The night's milk is strained into pans till morning, when the cream that will have risen is taken off, and the milk warmed to blood heat, when the cream is again returned to the milk and thoroughly mixed. This prevents the melting of the cream that would otherwise run off with the whey. The whole is then immediately laded into a tub with the morning's milk, and set for the cheese, with rennet sufficient to form the curd in about thirty minutes ; and here much care is thought to be necessary in cutting and crossing the curd, and much moderation in dipping and draining the whey from it, that the white whey (so called) may not exude from it.

"When sufficiently drained, it is taken and cut with a sharp knife to about the size and form of dice, when it is salted with about one pound of fine salt to twenty-five of curd. It is then subjected to a moderate pressure at first, gradually increasing it for two days, in the mean time turning it twice a day, and substituting dry cloths. It is then taken from the press and dressed all over with hot melted butter, and covered with thin cotton cloth, and this saturated with the melted butter. It is then placed upon the shelf, and turned and rubbed daily with the dressing until ripe for use."

One of the most important processes in the manufacture of good cheese is the preparation of the rennet. This is made of the inner lining or mucous membrane of the stomach of the young sucking calf, sometimes called the bag or maw; and the use of it was undoubtedly suggested, originally, by observing the complete and rapid coagulation or curdling of milk in the stomach of a calf newly killed. "Coagulation is the first process of digestion in the fourth stomach of the calf. There are numerous glands scattered in and about the stomach that secrete a fluid which readily and almost immediately accomplishes this coagulation. They are always full of it; even after the animal is dead they remain filled with it; and if the stomach is preserved from putrefaction, this fluid retains its coagulating quality for a considerable period; therefore dairy-women usually take care of the maw or stomach of the calf, and pre-

serve it by salting it, and then, by steeping it, or portions of it, in warm water, they prepare what they call
a rennet. After the maw has been salted a certain time,
it may be taken out and dried, and then it will retain
the same property for an indefinite period. A small
piece of the maw thus dried is steeped over night in a
few teaspoonfuls of warm water, and this water will
turn the milk of three or four cows."

It is important that rennet enough should be prepared at once for the whole season, in order to secure
as great a uniformity in strength as possible. The
object should be to produce a prompt, complete, and
firm or compact coägulation of all the cheesy matter.

Mr. Aiton, in his admirable treatise on the Dairy Husbandry of Scotland, gives the simple method of preparing the rennet in the dairy districts, as follows : " When
the stomach or bag — usually termed the yirning — is
taken from the calf's body, its contents are examined,
and if any straw or other food is found among the
curdled milk, such impurity is carefully removed ; but
all the curdled milk found in the bag is carefully preserved, and no part of the chyle is washed out. A
considerable quantity of salt — at least two handfuls —
is put into and outside the bag, which is then rolled up
and hung near a fire to dry. It is always allowed to
hang until it is well dried, and is understood to be
improved by hanging a year or longer before being
infused.

" When rennet is wanted, the yirning with its contents
is cut small, and put into a jar with a handful or two of
salt ; and a quantity of soft water that has been boiled
and cooled to sixty-five degrees, or of new whey taken
off the curd, is poured into it. The quantity of water
or whey necessary is more or less, according to the
quality of the yirning : if it is that of a new-dropped

calf, a Scotch chop pin, or at most three English pints, will be enough; but if the calf has been fed four or five weeks, two quarts or more may be used; the yirning of a calf four weeks old yields more rennet than that of one twice that age. When the infusion has remained in the jar from one to three days, the liquid is drawn off and strained, after which it is bottled for use; and if a dram-glass of any ardent spirit is put into each bottle, the infusion may either be used immediately, or kept as long as may be convenient."

The mode of preparing rennet in the dairy districts of this country is various; but that adopted by Mr. Fish, of Herkimer, New York, already quoted, is simple and easy of application. He says: " Various opinions exist as to the best mode of saving rennet, and that is generally adopted which, it is supposed, will curdle the most milk. I have no objection to any mode that will preserve its strength and flavor so that it will be smelled and *tasted* with good relish *when put into the milk.* Any composition not thus kept I deem unfit for use, as the coägulator is an essential agent in cheesing the curd, and sure to impart its own flavor.

" The rennet never should be taken from the calf till the excrement shows the animal to be in perfect health. It should be emptied of its contents, salted, and dried, without any scraping or rinsing, and kept dry for one year, when it will be fit for use. It should not be allowed to gather dampness, or its strength will evaporate. To prepare it for use, into ten gallons of water, blood warm, put ten rennets; churn or rub them often for twenty-four hours; then rub and press them to get the strength; stretch, salt, and dry them, as before. They will gain strength for a second use. Make the liquor as salt as it càn be made, strain and settle it, separate it from the sediment, if any, and it is fit for use.

Six lemons, two ounces of cloves, two ounces of cinnamon, and two ounces of common sage, are sometimes
added to the liquor, to preserve its flavor and quicken
its action.   If kept cool in a stone jar, it will keep
sweet any length of time desired, and a uniform strength
is secured while it lasts.   Stir it before dipping off.   To
set milk, take of it enough to curdle milk firm in forty
minutes ; squeeze or rub through a rag annatto enough
to make the curd a cream color, and stir it in with the
rennet."   It will be seen that he adopts the practice of
removing the contents of the stomach.   This, it appears
to me, is the best calculated to promote cleanliness and
purity, so important in making a good-flavored cheese.

But in Cheshire, so celebrated for its superior cheese,
the contents of the stomach are frequently salted by
themselves, and after being a short time exposed to the
air are fit for use ; while the well-known and highly-
esteemed Limburg cheese is mostly made with rennet
prepared as in Ayrshire, the curd being left in the
stomach, and both dried together.   The general opinion
is that rennet, as usually prepared, is not fit to use till
nearly a year old.

Perhaps the plan of making a liquid rennet from new
and fresh stomachs, and keeping it in bottles corked
tight till wanted for use, would tend still further to
secure this end.

The use of annatto to color the cheese artificially is
somewhat common in this country, though probably not
so much so as in many other countries.   Annatto, or
annotto, is made from the red pulp of the seeds of an
evergreen tree of the same name, found in the West
Indies and in Brazil, by bruising and obtaining a precip-
itate.   A variety is made in Cayenne, which comes into
the market in cakes of two or three pounds.   It is bright
yellow, rather soft to the touch, but of considerable

solidity. The quantity used is rarely more than an
ounce to one hundred pounds, and the effect is simply
to give the high coloring so common to the Gloucester
and Cheshire cheeses, and to many made in this country. This artificial coloring is continued from an
idle prejudice, somewhat troublesome to the dairyman,
expensive to the consumer, and adding nothing to the
taste or flavor of the article. The annatto itself is so
universally and so largely adulterated, often by poisonous substances, such as lead and mercury, that the practice of using it by the cheese-maker, and of requiring the
high coloring by the consumer, might well be discontinued. The common mode of application is to dissolve it

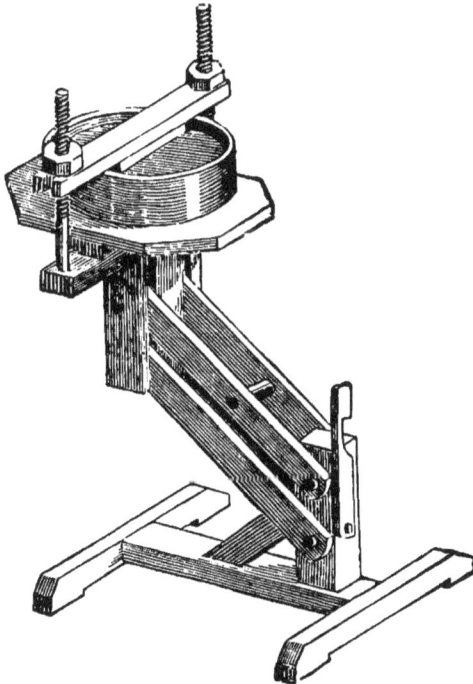

Fig. 82.   Cheese-press.

in hot milk, and add at the time of putting in the rennet,
or to put it upon the outside, in the manner of paint.
The cheese-presses in most common use are very dif-

ferent in construction, and each possesses. doubtless, some peculiar merits. The self-acting press, Fig. 82, is the favorite of some. Another form of this is seen in Fig. 83.

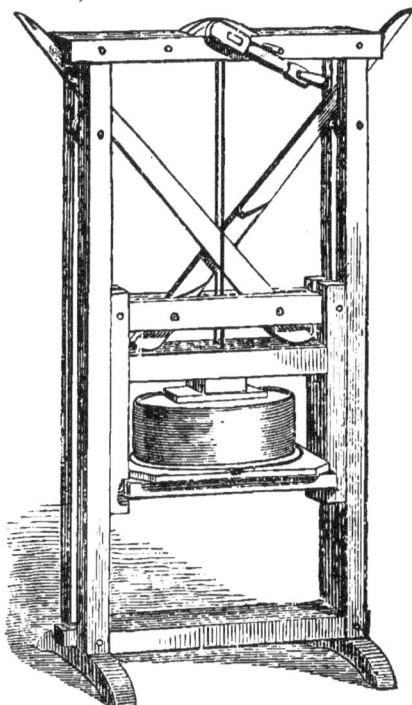

Fig. 83.   Self-acting cheese-press.

One of the most extensive and experienced dealers in cheese, in one of the largest dairy districts of New York, — Mr. Harry Burrill, of Little Falls, — has placed in my hands the following simple directions for cheese-making.

The cheese-tub should be so graduated that it may be correctly known what quantity of milk is used. This is requisite, in order that the proper proportions, both of coloring matter and rennet, may be used. The temperature should be ascertained by the thermometer. Experience proves that when the dairy has been at seventy degrees the best temperature at which to run the milk will be eighty-four degrees; but, as the temperature of the dairy at different times of the year will be found to vary above or below seventy degrees, the temperature of the milk must be proportionally regulated by the simple addition of cold water, to lower it; but, to increase the temperature, heat the milk in the usual manner, although it is absolutely necessary to avoid heating it beyond one hundred and twenty degrees.

After having brought the milk to the required temperature, and added the coloring, for every quarter hundred weight of cheese mix one pint of new sour whey with the requisite proportion of rennet; and, having arrived at the formation of a good curd, which will be the invariable result of a strict adhesion to the foregoing rules, let it be carefully cut up with three-bladed knives, as fine as possible ; then dip off half the whey, and heat a portion of it to the temperature of ninety-five degrees, and return it to the whey and curds; then, after stirring it for five minutes, allow the curd to sink, and as quickly as possible dip off the whey. Having done this, press the curd by placing on it a board weighted with from three to five fifty-pound weights, which will gradually and effectually press the remainder of the whey out.

When the whey is dipped off, put the curd into white twig basket-vats, made the shape and size of a turned vat, which would contain the sixth of a hundred weight (about three inches deep, and two feet in diameter). It will be necessary to have boards about one inch thick, and two feet four inches in diameter, to go between each of these twig vats, to prevent the whey running from one vat into the other. When it has been pressed, return it again into the cheese-tub, cut it into small pieces, put it into the vats again in dry cloths, press it and return it to the tub again, cutting it into small

22

pieces, and to every hundred weight of curd add one and one quarter pounds of salt; grind it twice, and stir it so that it shall be properly mixed with the salt; then put it into well-perforated turned vats, taking care to press it thoroughly whilst the vats are filling, to prevent the accumulation of air, to the presence of which is to be attributed the honeycomb appearance so often observed in cheese when cut.

When the cheese is put into the press let the pressure gradually upon it. After it has been in press one and a half hours, take it out and examine it, and, should there be any curd pressed over, cut it round and put it into the middle of the cheese, carefully breaking it up in the middle. Wash the ends of the cloths out in a bowl of warm water, squeeze them, and cover the cheese up, and, if there should be any not sufficiently full, it will be necessary either to put a follower upon it, or to put it into a smaller vat; in the evening let them be dry clothed. The following morning salt them all over and dry cloth them, and repeat this three successive mornings; after which, put them in vats, placed one on the other, and allow them to stand, if possible, a fortnight, occasionally wiping them. The cheese will get matured much sooner by these means, and the tendency to cracking and bulging be prevented.

The way to get a fine coat upon cheese, after the first coat has been washed and scraped off, is to put the cheese on shelves, nail thick sheeting to the ceiling from one of the shelves to the other, and let it drop closely to the floor. If put over the floor, cover them over with thick sheeting, or rugs.

The varieties of cheese are almost infinite in number, and are often dependent on very minute details of practice. The general principles involved are the same in all; but it would be next to impossible to find any one variety of cheese possessing uniformity throughout, in point of texture, consistency, taste, flavor, and keeping qualities; and it is rare, with the present guesswork in many of the operations of cheese-making, to find a lot of cheese made in the same dairy, from the same cows, on the same pastures, and by the same hands, which can be considered a fair sample of what is generally produced. These great differences are due to feeding and treatment of the cows in part, but especially to the temperature of the milk at the time of curding, which is again in part dependent on the quality and strength of the rennet employed.

Nothing is more susceptible to external influences, as has been remarked elsewhere, than milk and cream, both of which are liable to taint from the food of the cows, from impurities derived from careless milking, from exposure to foul or impure air in the cellar or milk-room, and from sudden changes in the atmosphere. The most scrupulous cleanliness is, therefore, required to produce a first quality of cheese, even under favorable circumstances. And when it is considered that it is necessary to observe minutely the temperature of the milk, and that slight differences at the time of forming the curd may make the difference of mellowness or toughness in the ripened cheese, and that the proper temperature is affected by the time taken to bring the curd, which depends on the strength and quality of the rennet, some of which will act in fifteen or twenty minutes, while the same quantity of others requires even two or three hours to produce the same effect, the infinite variety in the qualities of cheese will scarcely be a matter of surprise.

A brief statement of the mode of making some of the more important and well-known varieties will be sufficient in this connection. The details of cheese-making

in some of the best of the dairies of New England and
New York correspond in a remarkable degree with the
mode of making Cheddar and Cheshire cheese, both
celebrated for their richness and popularity in the market. Of the latter there are made, it is said, over
twelve thousand tons annually; Cheshire taking the
lead in cheese-making, and keeping about forty thousand
cows.

CHESHIRE CHEESE is remarkable for its uniformity,
being, in dairies of the best repute, made by fixed rules,
and usually by the same persons. If the number of
cows is sufficient to make a cheese from one meal, that
amount is used; if not, two meals are united. The
cows are milked at six o'clock, morning and evening;
are kept on rich pastures, and never driven far, great
care being taken that nothing shall interfere with the
regularity with which every operation connected with
this chief source of the wealth and prosperity of the
Cheshire farmer is conducted. The milk is brought in
large wooden pails into the milk-house, which it is generally contrived shall have a cool north aspect, and
immediately strained into pans, and placed upon the
floor of the dairy. Each pan is about six inches in
depth, and usually made of block-tin. This substance
is objected to by some because it is liable, like
every other metal, although, perhaps, in a less degree
than either zinc or lead, to be acted upon by the lactic
acid, and so produce compounds of a deleterious character. At six o'clock in the morning the cheese-ladder
is put on the cheese-tub, the whole of the night's milk
is again passed through the sieve, and the morning's
milk is then poured upon it, and well agitated to equal-
ize the temperature; in cold weather a pan of hot water
is previously put into the tub, to increase the temperature of the previous night's meal.

The rennet is next applied, care being taken that the heat of the whole quantity of the milk is about seventy-four degrees; and, almost simultaneously with the rennet, the annatto, — about a quarter of an ounce is sufficient for a cheese of sixty-four pounds, — both of which, in all well-regulated dairies, are strained through a piece of silk or fine cloth. The rennet is generally made on the previous evening, by a piece of the dried skin about the size of a crown-piece being immersed in hot water, and allowed to stand all night. After the rennet and coloring matter have been thoroughly mixed with the milk, it is covered with the lid of the cheese-tub, and in cold weather with a cloth in addition, to preserve the temperature of the mass until the curd has formed. It is then left undisturbed for about an hour, and frequently longer, to allow the coägulation of the milk. After that time a curd-breaker is passed up and down it for about five minutes, and again it is allowed to settle for another half-hour. The whey is then taken out by means of a dish or bowl, the curd being gathered to one side of the tub, and gently pressed by the hand, to allow the whey to separate from it more easily. It is then pressed by a weight of about fifty pounds; afterwards the curd is taken out of the tub and put into a basket, the inside of which is covered with a coarse square cheese-cloth. The four ends of the cloth are then folded over the curd, a tin hoop being put around the upper edge of the cheese, and within the sides of the vat, upon which a board is placed bearing a weight of about one hundred pounds, varying, of course, with the size of the cheese. This process is repeated two or three times, the curd being slightly broken at each operation. It is next taken out of the basket for salting or curing, and either broken down small by hand or in a curd-mill. A certain quantity of

22*　　　　17

salt is then carefully and intimately mixed with the
curd, according to the experience, taste, and custom,
of the dairymaid. It is then put into the cheese-vat in
a coarse cloth, pressed lightly at first for an hour; then
taken out and turned, and the pressure increased until
the proper degree of consistence is attained. Afterwards it is turned every twelve hours for three or four
days, remaining in the vat until the curd becomes so
dry as not to moisten the cloth. During this time
skewers are passed through holes made in the sides
of the vat into the body of the cheese, the more effectually to aid the expression of the whey, the pressure
being still continued. When they are withdrawn, the
whey flows through these miniature tunnels, which are
in a few moments obliterated by the superincumbent
weight.

It is the practice of some dairymaids in this county
to take the cheese to a cool salting-house, leaving it
there for a week or ten days, turning it daily, and rubbing salt on the upper surface. Others immerse the
cheese in a brine almost sufficiently strong to float it,
with occasional turning; others, again, after taking the
cheese from the press, place it in a furnace at a moderate heat, and keep it closed therein for a night; while
some run a hot iron over the whole, or over the edges.
The binder — a cloth of three or four inches in breadth
— is then passed tightly round the cheese, and secured
by pins, when it is removed to the cheese-room, and
placed on a kind of grass, which in Cheshire is called
sniggle, the newest or latest-made cheese being put in
the warmest situation. Here it remains, being turned
over three times a week while it is new, and less often
as it becomes matured, care being taken to keep each
one of the cheeses separate from all the others. The
room selected for a store is always that which can be

best protected from the light, and any sudden changes of temperature. The best Cheshire cheese is seldom ripe for the market under one or two years.

The STILTON CHEESE is by far the richest of the English dairies. This originated in a small town of that name, in Leicestershire. It possesses "a peculiar delicacy of flavor, a delicious mellowness, and a great aptness to acquire a species of artificial decay; without which, to the somewhat vitiated taste of lovers of Stilton cheese, as now eaten, it is not considered of prime account. To be in good order, according to the present standard of taste, it must be decayed, blue, and moist." To suit this taste, an artificial mode is adopted, old and decayed cheese being introduced into the new, or port wine or ale added by means of tasters, or caulking-pins are stuck into them, and left till they rust and produce an appearance of decay in the cheese.

"It is commonly made by putting the night's cream to the milk of the following morning with the rennet, great care being taken that the milk and the cream are thoroughly mixed together, and that they both have the proper temperature. *The rennet should also be very pure and sweet.* As soon as the milk is curdled, the whole of it is taken out, put into a sieve gradually to drain, and moderately pressed. It is then put into a case or box, of the form that it is intended to be; for, on account of its richness, it would separate and fall to pieces were not this precaution adopted. Afterwards it is turned every day on dry boards, cloth-binders being tied around it, which are gradually tightened as occasion requires. After it is removed from the box or hoop, the cheese must be closely bound with cloths and changed daily, until it becomes sufficiently compact to support itself. When these cloths are taken away, each cheese has to be rubbed over with a brush once every day. If the weather is moist or damp, this is done twice a day during two or three months. It is occasionally powdered with flour, and plunged into hot water. This hardens the outer coat and favors the internal fermentation, and thus produces what is called the ripening of the cheese. Sometimes it is made in a net like a cabbage-net, which gives it the form of an acorn."

The maturity of Stilton cheeses is sometimes hastened by putting them in a bucket, and covering them over with horse-dung.

GLOUCESTER CHEESE is likewise quite celebrated for its richness, piquancy, and delicacy of flavor, and justly commands a high price in the market. The management of the milk up to the time of curding is similar to that of Cheshire; a cheese, often being made of one meal, requires no additional heat to raise it to a proper temperature. After the curd is cut into small squares, the whey is carefully drained off through a hair strainer. The cutting is repeated every thirty minutes till the whey is removed, when it is put into vats and covered with dry cloths, and placed in the press. After remaining a sufficient length of time, it is put into a curd-mill and cut or ground into small pieces, when it is again packed in fine canvas cloth, and put in the cheese-vat. Hot water or whey is poured over the cloth, to harden the rind and prevent its cracking. " The curd is next turned out of the vat into the cloth, and, the inside of the vat being washed with whey, the inverted curd with the cloth is returned to the vat. The cloth is then folded over, and the vat put into the press for two hours, when it is taken out, and dry cloths applied during the course of the day. It is then replaced in the press until salted, which operation is generally performed about twenty-four hours after it is made. In salting the cheese, it is rubbed with finely-powdered salt, and this is thought to make the cheese more smooth and solid than when the salting process is performed upon the curd. The cheese is after this returned to the vat, and put under the press, in which several are placed, the newest at the bottom and the oldest on the top. The salting is repeated three times, twenty-four hours being allowed to intervene between each; and the cheese is finally taken from the press to the cheese-room in the course of five days. In the cheese-room it is turned over every day for a month, when it is cleaned of all scurf, and rubbed over with a woollen cloth dipped in a paint made of Indian red or Spanish brown and small beer. As soon as the paint is dry, the cheese is rubbed once a week with a cloth. The quantity of salt employed is about three and a half pounds; and one pound of annatto is sufficient to color half a ton of cheese."

CHEDDAR CHEESE is another variety in high repute for its richness, and commands a high price in the market. It is made of new milk only, and contains more fat than the egg. It is, indeed, too rich for ordinary consumption. The milk is set with rennet while yet warm, and allowed to stand still about two hours. The whey first taken off is heated and poured back upon the curd, and, after turning off the remainder, that is also heated and poured back in the same manner, where it stands about half an hour. The curd is then put into the press, and treated very much as the Cheshire up to the time of ripeness.

The DUNLOP CHEESE, the most celebrated of Scotland, had its origin in Ayrshire, from which it was sent to the Glasgow market, and from which the manufacture soon spread to Lanark, Renfrew, and other adjoining counties. It is manufactured, according to Aiton, in the following manner: When the cows on a farm are not so numerous as to yield milk sufficient to make a cheese every time they are milked, the milk is stored about six or eight inches deep in the coolers, and placed in the milk-house until as much is collected as will form a cheese of a proper size. When the cheese is to be made, the cream is skimmed from the milk in the coolers, and, without being heated, is, with the milk that is drawn from the cows at the time, passed through the sieve into the curd-vat. The cold milk from which the cream has been taken is heated so as to raise the temperature of the whole mass to near blood heat; and the whole is coagulated by the means of rennet carefully mixed with the milk. The cream is put into the curd-vat, that its oily parts may not be melted, and the skimmed milk is heated sufficient to raise the whole to near animal heat.

It may be said that the utmost care is always taken to keep the milk, in all stages of the operation, free not only from every admixture or impurity, but also from being hurt by foul air arising from acidity in any milky substance, putrid water, the stench of the barn, dunghill, or any other substance; and likewise to prevent the milk from becoming sour, which, when it happens, greatly injures the cheese. Great care is taken to prevent any of the butyraceous or oily matter in the cream from being melted in any stage of the process. To cool the milk, and to facilitate the separation or rising of the cream, a small quantity of clean cold water is generally mixed with the milk in each cooler. The coagulum is formed in from ten to fifteen minutes, and nobody would use rennet twice that required more than twenty minutes or half an hour to form a curd. Whenever the milk is completely coagulated the curd is broken, in order to let the serum or whey be separated and taken off. Some break the curd slightly at first, by making cross-scores with a knife or a thin piece of wood, at about one or two inches distant, and intersecting each other at right angles; and these are renewed still more closely after some of the whey has been discharged. Others break the whole curd more minutely at once with the hand or the skimmer.

After the curd has been broken, the whey ought to be taken off as speedily as it can be done, and with as little further breaking or handling the curd as possible. It is necessary, however, to turn the curd, cut it with a knife, or break it gently with the hand.

When the curd has consolidated a little, it is cut with the cheese-knife, slightly at first, and more minutely as it hardens, so as to bring off the whey. When the greater part of the whey has been extracted, the curd is taken up from the curd-boyn, and, being cut into pieces of about two inches in thickness, it is placed in a sort of vat or sieve with many holes. A lid is placed upon it, and a slight pressure, say from three to four stone avoirdupois; and the curd is turned up and cut small every ten or fifteen minutes, and occasionally pressed with the hand so long as it continues to discharge serum. When no more whey can be drawn off by these means, the curd is cut as small as possible with the knife, the proper quantity of salt minutely mixed into it in the curd-boyn, and placed in the chessart within a shift of thin canvas, and put under the press.

All these operations ought to be carried on and completed with the least possible delay, and yet without precipitation. The sooner the whey is removed after the coägulation of the milk, so much the better. But, if the curd is soft, from being set too cold, it requires more time, and to be more gently dealt with, as otherwise much of the curd and of the fat would go off with the whey; and when the curd has been formed too hot,.

the same caution is necessary. Precipitation, or handling the curd too roughly, would add to its toughness,
and expel still more of the oily matter ; and, as has been
already mentioned, hot water or whey should be put on
the curd when it is soft and cold, and cold water when
the curd is set too hot.

Undue delay, however, in any of these operations,
from the time the milk is taken out of the coolers until
the curd is under the press in the shape of a cheese,
is most improper, as the curd in all these stages is,
when neglected for even a few minutes, very apt to
become ill-flavored.   If it is allowed to remain too
long in the curd-vat, or in the dripper over it, before
the whey is completely extracted, the curd becomes
too cold, and acquires a pungent or acrid taste : or, it
softens so much that the cheese is not sufficiently adhe-
sive, and does not easily part with the serum. Whenever
the curd is completely set, the whey should be taken off
without delay ; and the dairymaid should never leave
the curd-boyn until the curd is ready for the dripper or
cheese-vat.   The salt is mixed into the curd.

After the cheese is put into the press, it remains for
the first time about an hour, or less than two hours,
until it is taken out, turned upside down in the cheese-
vat, and a new cloth put around it every four or six
hours until the cheese is completed, which is generally
in the course of a day and a half, two, or at most in
three days after it was first put under the press.

Some have shortened the process of pressing by
placing the cheese — after it has been under the press
for two hours or so for the first time — into water
heated to about one hundred or one hundred and ten
degrees, and allowing the cheese to remain in the water
about the space of half an hour, and thereafter drying
it with a cloth, and putting it again under the press.

When taken from the press, generally after two or three days from the time they were first placed under it, they are exposed for a week or so to the warmth and heat of the farmer's kitchen, — not to excite sweating, but merely to dry them a little before they are placed in the store, where a small proportion of heat is admitted. While they remain in the kitchen they are turned over three or four times every day ; and, whenever they begin to harden a little on the outside, they are laid up on the shelves of the store, where they are turned over once a day or once in two days for a week or so, until they are dry, and twice every week afterwards.

The store-houses for cheese in Scotland are in proportion to the size of the dairy, — generally a small place adjoining the milk-house, or in the end of the barn or other buildings, where racks are placed, with as many shelves as can hold the cheeses made in the season. When no particular place is prepared, the racks are placed in the barn, which is generally empty during summer; or some lay the cheeses on the floor of a garret over some part of their dwelling-house.

Wherever the cheeses are stored, they are not sweated or put into a warm place, but kept cool, in a place in a medium state, between damp and dry, without the sun being allowed to shine on them, or yet a great current of air admitted. Too much air, or the rays of the sun, would dry the cheeses too fast, diminish their weight, and make them crack ; and heat would make them sweat or perspire, which extracts the fat, and tends to induce hooving. But when they are kept in a temperature nearly similar to that of a barn, the doors of which are not much open, and but a moderate current of air admitted, the cheeses are kept in a proper shape, — neither so dry as to rend the skin, nor so

23

damp as to render them mouldy on the outside; and no
partial fermentation is excited, but the cheese is preserved sound and good.

DUTCH CHEESE. — The most celebrated of the Dutch
cheeses is the Edam, of North Holland, and the Gouda.
The manufacture of these and other varieties will be
described in a subsequent chapter, on Dairy Husbandry
in Holland.

The PARMESAN is an Italian cheese, made of one meal
of milk, allowed to stand sixteen hours, to which is
added another which has stood eight hours. The cream
being taken from both, the skim-milk is heated an hour
over a slow fire, and constantly stirred till it reaches
about eighty-two degrees, when the rennet is put in and
an hour allowed to form the curd. The curd is
thoroughly broken or cut, after which a part of the
whey is removed, and the curd is then heated nearly
up to the boiling point, when a little saffron is added to
color it. It then stands over the fire about half an
hour, when it is taken off, and nearly all the rest of
the whey removed, cold water being added, till the
curd is cool enough to handle. It is then surrounded
with a cloth, and, after being partially dried, is put
into a hoop and remains there two days. It is then
sprinkled with salt for thirty days in summer, or about
forty in winter. One cheese is then laid above another
to allow them to take the salt; after which they are
scraped and cleansed every day, and rubbed with linseed-oil to preserve them from the attack of insects, and
they are ready for sale at the age of six months.

AMERICAN CHEESE, as it is called in the English
markets, whither large quantities are shipped for sale,
is made of almost every conceivable variety and quality,
from the richest Cheddar or Cheshire to the poorest
skim-milk cheese. The statements of some of the best

dairymen have already been given. As a further illustration of the mode pursued in other sections of the
country, the statement of C. G. Taylor, a successful
competitor for the premiums offered by the Illinois State
Agricultural Society, may be given as follows :

" As the milk is drawn from the cows, it is immediately strained into a vat. This vat is a new patent, and
is better than any I have ever seen for cheese-making.
It is double, a space being left between the two parts.
Into the upper vat the milk is strained, and cold water
is applied between it and the lower one. Thus the animal heat is soon expelled, and the milk is prevented
from souring before morning. The morning milk is
added. Under the lower vat a copper boiler is
arranged. The water in the boiler is in perfect connection with that remaining all around the upper or
milk vat, connected with three copper pipes. With a
little wood the water is warmed. Thus the temperature of the milk is soon brought to the desired point to
receive the rennet, which is about ninety to ninety-
five degrees. Sufficient rennet is applied to the milk
to cause it to curdle or coägulate in from thirty to
forty minutes. Then the curd is carefully cut, each
way, into slices of about one inch square. Soon the
temperature is slowly increased. In about twenty
minutes the curd is carefully broken up with the hand,
— increasing the heat, and stirring often. When the
curd is sufficiently hard, so as to "*squeal*" when you bite
it, it is scalded. By this time the temperature is up to
about one hundred and thirty or one hundred and
forty.

" There are hinges placed in the legs of one end of
the vat, which is easily tipped, and through the curd-
strainer and whey-gate the whey is soon run off. The
curd is then dipped into a sink, over which is placed a

coarse strainer, and allowed to drain quite dry. It is then broken up fine, and one teacup of ground solar salt added to curd to make twenty pounds of cheese, and well worked in. After the curd is quite cool, it is placed in the hoop, and a light pressure is applied. In a few minutes more power is needed. After remaining in press about six hours, it is taken out of the hoop, wholly covered with strong muslin, finely sewed on, and then reversed and replaced in the hoop and press. It is allowed to remain until the next day, when it has to give place for another.

"After pressing thus twenty-four hours, the cheese is placed upon the shelf, and allowed to stand until the cloth is dry. Then a preparation, made from annatto and butter-oil, is applied sufficiently to fill all the interstices of the cloth. It must be turned and thoroughly rubbed three times a week, until ripe for use.

"I use the self-acting press. I know of none in use that is better, — the weight of the cheese being the power."

The statements of skilful and practical dairymen, in different parts of the country, are sufficient to show that good cheese can be produced; but it is believed that a more general attention to all the details of the dairy would add many thousand dollars a year to the wealth of the people, and enable us to compete successfully with the best dairy countries in the world.

The composition of cheese will, of course, differ widely in nutritive value, according to the mode of manufacture, age, etc. A specimen of good cheese was found to contain about 31.02 per cent. of flesh-forming substances, 25.30 per cent. of heat-producing substances, 4.90 per cent. of mineral matter, and 38.78 per cent. of water.

The analyses of several varieties will serve as a com

parison of cheese with other kinds of food. The Cheddar was a rich cheese two years old, the double Gloucester one year old, the Dunlop one year old, the skim-
milk one year.

	Cheddar.	Dbl. Glo'ster.	Dunlop.	Skim-milk.
Water, ...	30.04	35.81	38.46	43.82
Caseine, ...	28.98	37.96	25.87	45.04
Fat, ...	30.40	21.97	31.86	5.98
Ash, ...	4.58	4.25	8.81	5.18

Professor Johnston gives a table of comparison of
Cheddar and skim-milk cheese in a dried state, and milk,
beef, and eggs, also in a dried state, as follows :

	Milk.	Cheddar cheese, dried.	Skim-milk cheese, dried.	Beef.	Eggs.
Caseine (curd), .	35	45	80	89	55
Fat (butter), . .	24	48	11	7	40
Sugar, ...	37	–	–	–	–
Mineral matter, .	4	7	9	4	5
	100	100	100	100	100

A full-milk cheese differs but little from pure milk,
except in the absence of sugar, which, as already seen,
is held in solution, and goes off in the whey. The difference becomes greater in proportion as the cream is
removed from the milk before curding, and the nutritive
qualities thereby diminished.

Cheese is used both as a regular article of food, for
which the ordinary kinds of full-milk cheeses are
admirably fitted, and as a condiment or digester, in connection with other articles of food; and for this purpose
the stronger varieties, such as are partially decayed
and mouldy, are best. "When the curd of milk is
exposed to the air in a moist state, for a few days, at a
moderate temperature, it begins gradually to decay, to
emit a disagreeable odor, and to ferment. When in

23*

this state, it possesses the property, in certain circumstances, of inducing a species of chemical change and
fermentation in other moist substances with which it is
mixed, or is brought into contact. It acts after the
same manner as sour leaven does when mixed with
sweet dough. Now, old and partially decayed cheese
acts in a similar way when introduced into the stomach.
It causes chemical changes gradually to commence
among the particles of the food which has previously
been eaten, and thus facilitates the dissolution which
necessarily precedes digestion. It is only some kinds
of cheese, however, which will effect this purpose.
Those are generally considered the best in which some
kind of cheese-mould has established itself. Hence,
the mere eating of a morsel of cheese after dinner does
not necessarily promote digestion. If too new, or of
improper quality, it will only add to the quantity of
food with which the stomach is probably already over-
loaded, and will have to await its turn for digestion by
the ordinary processes." This mouldiness and tendency
to decay, with its flavor and digestive quality, are
often communicated to new cheese by inoculation, or
insertion of a small portion of the old into the interior
of the new by means of the cheese-taster.

In studying attentively the practice of the most successful cheese-makers, I think it will be observed that
they are particularly careful about the preparation of
the rennet, and equally so about the details of pressing.
In my opinion, the point in which many American
cheese-makers fail of success is in hurrying the pressing. I think it will be found that the best cheese is
pressed two days, at least, and in many cases still
longer.

CHAPTER X.

DAIRY STOCK, properly fed and managed, is liable to few diseases in this country, notwithstanding the sudden changes to which our climate is subject. If pure air, pure water, a dry barn or pasture, and a frequent but gradual change of diet, when kept in the stall, are provided for milch cows, nature will generally remedy any derangements of the system which may occur, far better than art. Common sense is especially requisite in the treatment of stock, and that will very rarely dictate a resort to bleeding, boring the horns, cutting off the tail, and a thousand other equally absurd practices, too common even within the memory of men still living.

The diseases most to be dreaded are garget, puerperal or milk fever, and idiopathic or common fever, commonly called " horn ail," and often " tail ail."

GARGET is an inflammation of the internal substance of the udder. One or more of the teats, or whole sections of the udder, become enlarged and thickened, hot, tender, and painful. The milk coägulates in the bag, and causes inflammation where it is deposited, which is accompanied by fever. It most commonly occurs in young cows after calving, especially when in too high condition. The secretion of milk is very much lessened, and, in very bad cases, stopped altogether. Sometimes the milk is thick, and mixed with blood. Often, also, in severe cases, the hind extremities, as the hip-joint, hock, or fetlock, are swollen and inflamed to such an extent that the animal cannot rise. The simplest remedy, in mild cases, is to put the calf to its mother several times a day. This will remove the flow of milk, and often dispel the congestion.

Sometimes the udder is so much swollen that the cow will not permit the calf to suck. If the fever increases, the appetite declines, and rumination ceases. In this stage of the complaint, the advice of a scientific veterinary practitioner is required. A dose of purging medicine and frequent washing of the udder, in *mild* cases, are usually successful. The physic should consist of Epsom salts one pound, ginger half an ounce, nitrate of potassa half an ounce ; dissolved in a quart of boiling water ; then add a gill of molasses, and give to the cow lukewarm. Diet moderate ; that is, on bran, or if in summer green food. There are various medicines for the different forms and stages of garget, which, if the above medicine fails, can be properly prescribed only by a skilful veterinary practitioner.

It is important that the udder should be frequently examined, as matter may be forming, which should be immediately released. Various causes are assigned for this disease, such as exposure to cold and wet, or the want of proper care or attention in parturition.

An able writer, Mr. Youatt, says that hasty drying up a cow often gives rise to inflammation and indurations of the udder, difficult of removal. Sometimes a cow lies down upon and bruises the udder, and this is another cause. But a very frequent source, and one for which there can be no excuse, is the failure to milk a cow clean. The calf should be allowed to suck often, and the cow should be milked at least twice a day as clean as possible, while suffering from this complaint.

If the udder is hot and feverish, a wash may be used, consisting of eight ounces of vinegar and two ounces of camphoretted spirit; the whole well and thoroughly mixed, and applied just after milking, to be washed off in warm water before milking again.

In very bad cases, iodine has often been found most effectual. An iodine ointment may be prepared by taking one drachm of hydriodate of potash and an ounce of lard, and mixing them well together. A small portion of the mixture, from the size of a pigeon's egg, in limited inflammations, to twice that amount, is to be well rubbed into the swollen part, morning and night.

When milk forms in the bag before parturition, so as to cause a swelling of the udder, it should be milked away; and a neglect of this precaution often leads to violent attacks of garget.

Prevention is always better than cure. The reason most commonly given for letting the cow run dry for a month or two before calving is that after a long period of milking her system requires rest, and that she will give more milk and do better the coming season than if milked up to the time of calving.

This is all true, and a reason sufficient in itself for drying off the cow some weeks before parturition; but there is another important reason for the practice, which is that the mixture of the old milk with the new secretion is liable to end in an obstinate case of garget.

To prevent any ill effects from calving, the cow should not be suffered to get too fat, which high feeding after drying off might induce.

The period of gestation is about two hundred and eighty-four or two hundred and eighty-five days. But cows sometimes overrun their time, and have been

18

known to go three hundred and thirteen days, and even more; while they now and then fall short of it, and have been known to calve in two hundred and twenty days. If they go much over the average time, the calf will generally be a male. But cows are sometimes liable to slink their calves; and this usually takes place about the middle of their pregnancy. To avoid the evil consequences, so far as possible, they should be watched; and, if a cow is found to be uneasy and feverish, or wandering about away from the rest of the herd, and apparently longing for something she cannot get, she ought to be taken away from the others.

If a cow slinks her calf while in the pasture with others, they will be liable to be affected in the same way.

In many cases, physicking will quiet the cow's excitement in the condition above described, and prove of essential benefit. A dose of one pound of Epsom or Glauber's salts, and one ounce of ginger, mixed in a pint of thick gruel, should be given first, to be immediately followed by the salts, in a little thinner gruel.

When a cow once slinks her calf, there is great risk in breeding from her. She is liable to do the same again. 'But when the slinking is caused by sudden fright or over-exertion, or any offensive matter, such as blood or the dead carcasses of animals, this result is not so much to be feared.

But the cow, when about to calve, ought not to be disturbed by too constant watching. The natural presentation of the foetus is with the head lying upon the fore legs. If in this position, nature will generally do all. But, if the presentation is unnatural, and the labor has been long and ineffectual, some assistance is required. The hand, well greased, may be introduced, and the position of the calf changed; and, when in a proper position, a cord should be tied round the fore legs, just above the hoofs; but no effort should be made to draw out the calf till the natural throes are re· peated. If the nostril of the calf has protruded, and the position is then found to be unnatural, the head cannot be thrust back without destroying the life of the calf. The false position most usually presented is that of the head first, with the legs doubled under the belly. A cord is then fixed around the lower jaw, when it is pushed back, to give an opportunity to adjust the fore legs, if possible. The object must now be to save the life of the cow.

But the cases of false presentation, though compara· tively rare, are so varied that no directions could be given which would be applicable in all cases.

After calving the cow will require but little care, if she is in the barn, and protected from changes of weather. A warm bran mash is usually given, and the state of the udder examined.

PUERPERAL OR MILK FEVER.— Calving is often at· tended with feverish excitement. The change of power· ful action from the womb to the udder causes much constitutional disturbance and local inflammation. A cow is subject to nervousness in such circumstances, which sometimes extends to the whole system, and causes puerperal fever. This complaint is called *dropping* after calving, because it succeeds that process. The prominent symptom is a loss of power over the motion of the hind extremities, and inability to stand; some· times loss of sensibility in these parts, so that a deep puncture with a pin, or other sharp instrument, is unfelt.

This disease is much to be dreaded by the farmer, on account of the high state of excitement and the local inflammation. Either from neglect or ignorance, the mal· ady is not discovered until the manageable symptoms have passed, and extreme debility has appeared. The animal is often first seen lying down, unable to rise; prostration of strength and violent fever are brought on by inflammation of the womb. But soon a general inflammatory action succeeds, rapid and violent, with complete prostration of all the vital forces, bidding defiance to the best-selected remedies.

Cows in very high condition, and cattle removed from low keeping to high feeding, are the most liable to puerperal fever. It occurs most frequently during the hot weather of summer, and then it is most dangerous. When it occurs in winter, cows sometimes recover. In hot weather they usually die.

Milk fever may be induced by the hot drinks often given after calving. A young cow at her first calving is rarely attacked with it. Great milkers are most commonly subject to it; but all cows have generally more or less fever at calving. A little addition to it, by improper treatment or neglect, will prevent the secretion of milk; and thus the milk, being thrown back into the system, will increase the inflammation.

This disease sometimes shows itself in the short space of two or three hours after calving, but often not under two or three days. If four or five days have passed, the cow may generally be considered safe. The earliest symptoms of this disease are as follows:

The animal is restless, frequently shifting her position; occasionally pawing and heaving at the flanks. Muzzle hot and dry, the mouth open, and tongue out at one side; countenance wild; eyes staring. She moans often, and soon becomes very irritable. Delirium follows; she grates her teeth, foams at the mouth, tosses her head about, and frequently injures herself. From the first, the udder is hot, enlarged, and tender; and if this swelling is attended by a suspension of milk, the cause is clear. As the case is inflammatory, its treatment must be in accordance; and it is usually subdued without much difficulty. Mr. Youatt says, · " The animal should be bled, and the quantity regulated by the impression made upon the circulation, — from six to ten quarts often before the desired effect is produced." He wrote at a time when bleeding was adopted as the universal cure, and before the general reasoning and treatment of diseases of the human system was applied to similar diseases of animals. The cases are very rare, indeed, where the physician of the present day finds it necessary to bleed in diseases of the human subject; and they are equally rare, I apprehend, where it is really necessary or judicious to bleed for the diseases of animals. A more humane and equally effectual course will be the following:

A pound to one and a half pounds of Epsom or Glauber's salts, according to the size and condition of the animal, should be given, dissolved in a quart of boiling water; and, when dissolved, add pulv. red pepper a quarter of an ounce, caraway do. do., ginger do. do.; mix, and add a gill of molasses, and give lukewarm. If this medicine does not act on the bowels, the quantity of ginger, capsicum, and caraway, must be doubled. The insensible stomach must be roused. When purging in an early stage is begun, the fever will more readily subside. After the operation of the medicine, sedatives may be given, if necessary.

The digestive function first fails, when the secondary or low state of fever comes on. The food undischarged ferments; the stomach and intestines are inflated with gas, and swell rapidly. The nervous system is also attacked, and the poor beast staggers. The hind extremities show the weakness; the cow falls, and cannot rise; her head is turned on one side, where it rests; her limbs are palsied. The treatment

24

in this stage must depend on the existence and degree of fever. The pulse will be the only true guide. If it is weak, wavering, and irregular, we must avoid depleting, purgative agents. The blood flows through the arteries, impelled by the action of the heart, and its pulsations can be very distinctly felt by pressing the finger upon almost any of these arteries that is not too thickly covered by fat or the cellular tissues of the skin, especially where it can be pressed upon some hard or bony substance beneath it. The most convenient place is directly at the back part of the lower jaw, where a large artery passes over the edge of the jaw-bone to ramify on the face. The natural pulse of a full-grown ox will vary from about forty-eight to fifty-five beats a minute; that of a cow is rather quicker, especially near the time of calving; and that of a calf is quicker than that of a cow. But a very much quicker rate than that indicated will show a feverish state, or inflammation; and a much slower pulsation indicates debility of some kind.

Next in importance, as we have already stated, is the physic. The bowels must be opened, or the animal will fall a victim to the disease. All medicines should be of an active character, and in sufficient quantity; and stimulants should always be added to the purgative medicines, to insure their operation. Ginger, gentian, caraway, or red pepper in powder, may be given with each dose of physic. Some give a powerful purgative, by means of Epsom salts one pound, flour of sulphur four ounces, powdered ginger a quarter of an ounce, all dissolved in a quart of cold water, and one half given twice a day till the bowels are opened The digestive organs are deranged in most forms of milk fever, and the third stomach is loaded with hard, indigestible food. When the medicine has operated, and the fever is subdued, little is required but good nursing to restore the patient.

No powerful medicines should be used without discretion; for in the milder forms of the disease, as the simple palsy of the hind extremities, the treatment, though of a similar character, should be less powerful, and every effort should be made for the comfort of the cow, by providing a thick bed of straw, and raising the fore quarters to assist the efforts of nature, while all filth should be promptly and carefully removed. She may be covered with a warm cloth, and warm gruel should be frequently offered to her, and light mashes. An attempt should be made several times a day to bring milk from the teats. The return of milk is an indication of speedy recovery.

Milch cows in too high condition appear to have a constitutional tendency to this complaint, and one attack of it predisposes them to another.

SIMPLE FEVER. — This may be considered as increased arterial action, with or without any local affection; or it may be the consequence of the sympathy of the system with the morbid condition of some particular part. The first is pure or idiopathic fever; the other, symptomatic fever. Pure fever is of frequent occurrence in cattle. Symptoms as follows: muzzle dry; rumination slow or entirely suspended; respiration slightly accelerated; the horn at the root hot, and its other extremity frequently cold; pulse quick; bowels constipated; coat staring, and the cow is usually seen separated from the rest of the herd. In slight attacks, a cathartic of salts, sulphur, and ginger, is sufficient. But, if the common fever is neglected, or improperly treated, it may assume, after a time, a local determination, as pleurisy, or inflammation of the lungs or bowels. In such cases the above remedy would be insufficient, and a veterinary surgeon, to manage the case, would be necessary. Symptomatic fever is more dangerous, and is commonly the result of injury, the neighboring parts sympathizing with the injured part. Cattle become unwell, are stinted in their feed, have a dose of physic, and in a few days are well; still, a fever may terminate in some local affection. But in both cases pure fever is the primary disease.

A more dangerous form of fever is that known as symptomatic. As we have said, cattle are not only subject to fever of common intensity, but to symptomatic fever, and thousands die annually from its effects. But the young and the most thriving are its victims. There are few premonitory symptoms of symptomatic fever. It often appears without any previous indications of illness. The animal stands with her neck extended, her eyes protruding and red, muzzle dry, nostrils expanded, breath hot, base of the horn hot, mouth open, pulse full, breathing quick. She is often moaning; rumination and appetite are suspended; she soon becomes more uneasy; changes her position often. Unless these symptoms are speedily removed, she dies in a few hours. The name of the ailment, inflammatory or symptomatic fever, shows the treatment necessary, which must commence with purging. Salts here, as in most inflammatory diseases, are the most reliable. From a pound to a pound and a half, with ginger and sulphur, is a dose, dissolved in warm water or thin gruel. If this does not operate in twelve hours, give half the dose, and repeat once in twelve hours, until the bowels are freed. After the operation of the medicine the animal is relieved. Then sedative medicines may be given. Sal ammoniac one drachm, powdered nitre two drachms, should be administered in thin gruel, two or three times a day, if required.

Typhus fever, common in some countries, is little known here among cattle.

TYPHOID FEVER sometimes follows intense inflammatory action, and is considered the second stage of it. This form of fever is usually attended with diarrhœa. It is a debilitating complaint, and is sometimes followed by diseases known as black tongue, black leg, or quarter evil. The cause of typhoid fever is involved in obscurity. It may be proper to say that copious drinks of oat-meal gruel, with tincture of red pepper, a diet of bran, warmth to the body, and pure air, are great essentials in the treatment of this disease.

The barbarous practices of boring the horns, cutting the tail, and others equally absurd, should at once and forever be discarded by every farmer and dairyman. Alternate heat or coldness of the horn is only a symptom of this and other fevers, and has nothing to do with their cause. The horns are not diseased any further than a determination of blood to the head causes a sympathetic heat, while an unnatural distribution of blood, from exposure or other cause, may make them cold.

In all cases of this kind, if anything is done, it should be an effort to assist nature to regulate the animal system, by rousing the digestive organs to their natural action, by a light food, or, if necessary, a mild purgative medicine, followed by light stimulants.

The principal purgative medicines in use for neat cattle are Epsom salts, linseed-oil, and sulphur. A pound of salts will ordinarily be sufficient to purge a full-grown cow.

A slight purgative drink is often very useful for cows soon after calving, particularly if feverish, and in cases of over-feeding, when the animal will often appear dull and feverish; but when the surfeiting is attended

24*

by loss of appetite, it can generally be cured by withholding food at first, and then feeding but slightly till
the system is renovated by dieting.

Purgative drinks will often cure cases of red water,
if taken in season.

A purgative is often necessary for cows after being
turned into a fresh and luxuriant pasture, when they
are apt to become bound from over-feeding; but constipation does not so often follow a change from dry to
green food in spring, as from a poor pasture in summer
to one where they obtain much better feed.

The HOOVE or HOVEN is brought on by a derangement of the digestive organs, occasioned by over-feeding on green and luxuriant clover, or other luxuriant
food. It is simply the distension of the first stomach by carbonic acid gas. In later stages, after fermentation of the contents of the stomach has commenced, hydrogen gas is also found. The green food,
being gathered very greedily after the animal has been
kept on dry and perhaps unpalatable hay, is not sent
forward so rapidly as it is received, and remains to
overload and clog the stomach, till this organ ceases or
loses the power to act upon it. Here it becomes moist
and heated, begins to ferment, and produces a gas
which distends the paunch of the animal, which often
swells up enormously. The cow is in great pain, breathing with difficulty, as if nearly suffocating. Then the body
grows cold, and, unless relief is at hand, the cow dies.

Prevention is both cheaper and safer than cure; but
if by neglect, or want of proper precaution, the animal
is found in this suffering condition, relief must be
afforded as soon as possible, or the result will be fatal.

A hollow flexible tube, introduced into the gullet,
will sometimes afford a temporary relief till other means
can be had, by allowing a part of the gas to escape;

but the cause is not removed either by this means or by puncturing the paunch, which is often dangerous.

In the early stage of the disease the gas may be neutralized by ammonia, which is usually near at hand. Two ounces of liquid ammonia, in a quart of distilled or rain water, given every quarter of an hour, will prove beneficial. A little tincture of ginger, essence of anise-seed, or some other cordial, may be added, without lessening the effect of the ammonia.

If the case has assumed an alarming character, the flexible tube, or probang, may be introduced, and afterwards take three drachms either of the chloride of lime or the chloride of soda, dissolve in a pint of water, and pour it down the throat. Lime-water, potash, and sulphuric ether, are often used with effect.

In desperate cases it may be found necessary to make an incision through the paunch; but the chloride of lime will, in most cases, give relief at once, by neutralizing the gas.

CHOKING is often produced by feeding on roots, particularly round and uncut roots, like the potato. The animal slavers at the mouth, tries to raise the obstruction from the throat, often groans, and appears to be in great pain. Then the belly begins to swell, from the - amount of gases in the paunch.

The obstruction, if not too large, can sometimes be thrust forward by introducing a flexible rod, or tube, into the throat. This method, if adopted, should be attended with great care and patience, or the tender parts will be injured. If the obstruction is low down, and a tube is to be inserted, a pint of olive or linseed oil first turned down will so lubricate the parts as to aid the operation, and the power applied must be steady. If the gullet is torn by the carelessness of the operator, or the roughness of the instrument, a rupture generally results in serious consequences. A hollow tube is best, and if the object is passed on into the paunch, the tube should remain a short time, to permit the gas to escape. In case the animal is very badly swelled, the dose of chloride of lime, or ammonia, should be given, as for the hoove, after the obstruction is removed.

Care should be taken, after the obstruction is removed, to allow no solid food for some days.

FOUL IN THE FOOT. — Cows and other stock, when fed in low, wet pastures, will often suffer from ulcers or sores, generally appearing first between the claws. This is commonly called foul in the foot, and is analogous to foot-rot in sheep. It is often very painful, causing severe lameness and loss of flesh, and discharges a putrid matter, or pus. Sometimes it first appears in the form of a swelling near the top of the hoof, which breaks and discharges foul matter.

The rough and common practice among farmers is to fasten the foot in the same manner as the foot of an ox is fastened in shoeing, and draw a rough rope back and forth over the ulcerated parts, so as to produce a clean, fresh wound, and then dress it with tar or other similar substance.

This is often an unnecessarily cruel operation. The loose matter may easily be removed by a knife, and then carefully wiped off with with a moist sponge. The animal should then be removed at once to a warm, dry pasture, or kept in the barn.

If the case has been neglected till the pasterns become swollen and tender, the sore may be thoroughly cleansed out, and dressed with an ointment of sulphate of iron one ounce, molasses four ounces, simmered over a slow fire till well mixed. Apply on a piece of cotton batting, and secure upon the parts. If any morbid growth or fungus appear, use equal parts of powdered blood-root and alum sprinkled on the sore, and this will usually effect a cure.

Some also give a dose of flour of sulphur half an ounce, powdered sassafras-bark one ounce, and burdock two ounces, the whole steeped in a quart of boiling water, and strained when cool; and, if the matter still continues to flow from the sore, wash it morning and night with chloride of soda one ounce, or a table-spoonful of common salt dissolved in a pint of water.

Foul in the foot causes very serious trouble, if not taken in season. The health of cows is injured to a great extent. I have seen, during the present season, many instances of foul in the foot in dairy stock arising from the wetness of the pastures. No lameness in cattle should be neglected.

RED WATER is so called from the high color of the urine. It is rather a symptom of some derangement of the digestive organs than a disease of itself, and the cause is most frequently to be found in the quality of the food. It is peculiar to certain localities, and is of very rare occurrence in New England.

In the early stage of the difficulty the bowels are loose, but soon constipation ensues, and the appetite is affected, the milk decreases, and the urine becomes either very red or sometimes black.

The case demands treatment, for it is apt to prey upon the health of the cow. Purgatives are usually employed with most success. Take a pound of Epsom salts, half an ounce of ginger, and half an ounce of carbonate of ammonia. Pour a quart of boiling water on the salts and ginger, stir thoroughly, and, when cold, add the ammonia. If this fails to act on the bowels, repeat a quarter part of it every six or eight hours till it succeeds. Then a nutritious diet should be used till the appetite is fully restored.

If a cow is once affected in this way, the difficulty will be liable to return, and she had better be disposed of.

HOOSE is a cold or cough to which stock are subject when exposed to wet weather and damp pastures.

The cold may not be bad at first, or may be so slight as not to attract attention; but it often leads to worse complaints, and ought, when observed, to be attended to at once, by keeping the animal in a dry and warm barn a few days, and feeding with mashes, and, if it continues, take an ounce of sweet spirits of nitre in a pint of ginger tea; mix, and give in a quart of thick gruel.

No prudent farmer will neglect to observe approaching symptoms of disease in his stock. The cheapest way to keep animals healthy is to treat them properly in time, and before disease is seated upon them. Hoose often ends in consumption and death.

INFLAMMATION OF THE GLANDS often occurs in hoose, catarrh, etc., but they resume their natural state when these complaints are removed. The animal cannot swallow without pain sometimes, and soft food should be given. Remove the cause, and the inflammation ceases. Some make a relaxing poultice of marsh-mallows, or similar substances; and rub the throat with a mixture of olive or goose oil one gill, spirit of camphor one ounce, oil of cedar one ounce, and half a gill of vinegar.

INFLAMMATION OF THE LUNGS.— Common catarrh or hoose sometimes leads to inflammation of the lungs, which is indicated by dulness and sore cough. The ears, the roots of the horns, and legs, are sometimes cold. The breath is hot, as well as the mouth; and the animal rarely lies down, and is reluctant to move, or change its position. Warm water and mashes, or gruel, may be given, and the animal kept in a dry place. The cause of the complaint should be removed, and the trouble will generally soon cease. The treatment is much the same as for fever; but where the surface of the body is cold, as is generally the case, give sweet spirits of nitre two ounces, liquor acetate of ammonia four ounces, in a pint of water, two or three times a day.

DIARRHŒA is brought on by too sudden change of food, especially from dry to green and succulent food; sometimes by poisonous plants or bad water. If slight, the farmer may not be anxious to check it. It may show simply an effort of nature to throw off some injurious substances from the body, and so it may exist when the animal is quite healthy. But, if it continues too long, and is likely to debilitate the system, a mild purgative may be given to assist rather than check the operation of nature. Half a pound of Epsom salts, with a little ginger and gentian, will do for a medium-sized animal in this case; but a purgative may be followed in a day or two by an astringent medicine. Take prepared chalk two ounces, powdered oak-bark one ounce, powdered catechu two drachms, powdered opium one drachm, and four drachms powdered ginger. Mix these together, and give in a quart of warm gruel. Sometimes a few ounces of pulverized charcoal will arrest the diarrhœa. Common diarrhœa may be distinguished from dysentery by a too abundant discharge of dung in too fluid a form, or in a full, almost liquid stream, sometimes very offensive to the smell, and now and then bloody. In dysentery, the dung is often mixed with mucus and blood, and is not unfrequently attended by a hard straining. The quantity of dung is less than in diarrhœa, but more offensive.

Diarrhœa may occur at any season of the year, and sometimes leads to dysentery, which more frequently appears in the spring and fall.

DYSENTERY, or scouring rot, is a dangerous and trouble-some malady when it becomes seated.

The cow suffers from painful efforts to pass the dung, which is thin, slimy, olive-colored, and offensive, and after it falls rises up in little bubbles, with a slimy substance upon it. She is restless, lying down and soon rising again, and appears to be in great distress. The hair seems to stand out stiff from the body, and this stage of the malady indicates an obstinate and fatal disease.

It is often brought on by a simple cold at the time of calving, exposure to sudden changes, and by poor keeping, which exhausts the system, especially in winter. A dry, warm barn, and careful nursing, will do much ; and dry, sweet food, as hay, oat-meal, boiled potatoes, gruel, &c. Some linseed-meal is also very good for cows with this complaint. A little gum-arabic or starch may be mixed with the medicine.

The treatment is much the same as for diarrhœa.

The MANGE is commonly brought on by half starving in winter, and by keeping the cow in a filthy, ill-ventilated place. It is contagious, and if one cow of a herd has it, the rest will be apt to get it also. Blaine says, " Mange has three origins, — filth, debility, and contagion." It is a disgrace to the farmer to suffer it to enter his herd from either of these causes, since it shows a culpable neglect of his stock. I am sorry to say it is too common in this country, especially in filthy barns.

The cow afflicted with the mange is hide-bound ; the hair is dry and stiff, and comes off. She is constantly rubbing, and a kind of white scurfiness appears on the skin. It is most perceptible towards the latter part of winter and in spring, and thus too plainly tells the story of the winter's neglect.

An ointment composed chiefly of sulphur has been found most effectual. Some mercurial ointment may be added, if the cows are kept housed; but, if let out during the day, the quantity must be very small, else salivation is produced by their licking themselves.

The ointment may be made of flour of sulphur one pound, strong mercurial ointment two ounces, common turpentine one half-pound, lard one and a quarter pounds. Melt the turpentine and lard together, and stir in the sulphur as they begin to cool off; then rub down the mercurial ointment on some hard substance with the other ingredients. Rub the whole in with the hand, and take care to leave no places untouched, once a day, for three days; and after this, if any places are left un-cured, rub it in over them. There is no danger in this application, if the animal is not exposed to severe cold. This will be pretty sure to effect a speedy cure, if aided by cleanliness, pure air, and a nutritious diet.

Another wash for mange is the following: Pyrolig-neous acid four ounces, water a pint; mix and apply.

LICE show unpardonable neglect of duty wherever they are suffered to exist. They crawl all over the stable-floor and the stalls, on the pastures, and a touch is sufficient to give them to other animals. They worry and trouble the poor animal constantly; and no thriftiness can be expected where they are found. If the mange ointment does not completely destroy them, as it often will, take bees-wax, tallow, and lard, in equal parts, and rub it into the hide in the most thorough man-ner, with the hand or a brush, two and a half pounds for a small cow, three pounds for a large one. The next day it may be washed off in soft soap, and the lice will have disappeared from the animal, but not always from the barn. Some use a wash of powdered lobelia-seeds two ounces, steeped in boiling water, and

applied with a sponge. Others hang up tobacco-leaves
over the stalls. This may do to keep them away ; but,
after the animal is covered with them, they are not so
easily scared.

WARBLES.— The gad-fly is very troublesome to cattle
towards the end of summer. The fly alights on the back
of the cow, punctures the skin, and lays her eggs under
it. A tumor is now formed, varying in size, which soon
bursts and leaves a small hole for the grub already
hatched to breathe through. Here the insect feeds on
its surroundings, and grows up to considerable size.
All this time the animal is probably suffering more or
less pain, and often tries to lick or rub the part affected,
if possible. Farmers often press them out with the finger and thumb. The best way is to puncture the skin
with a common pen-knife, and then press out the
grub. They injure the hide more than most people are
aware of.

LOSS OF CUD is a consequence of indigestion, and is
often brought on by eating too greedily of food which
the cow is not used to. Loss of cud and loss of appetite
are synonymous. Gentle purgatives may be given,
with such as salts, ginger, and sulphur. But when a cow
is surfeited, as already said, I should prefer to withhold
food entirely, or for the most part, till the system can
regulate itself.

DISEASES OF CALVES. — The colostrum, or first milk of
the cow after calving, contains medicinal qualities peculiarly adapted to cleanse the young calf, and free its
bowels from the matter always existing in them at birth.
This should, therefore, never be denied it. Bleeding
at the navel, with which calves are sometimes seriously
troubled, may generally and safely be stopped by tying
a string around the cord which hangs suspended
from it.

But DIARRHŒA, PURGING, or SCOURS, is the most dangerous complaint with which calves are afflicted. This is caused often by neglect, or exposure to wet and cold, or insufficiency of food at one time and over-feeding at another. Stinting the calf in food or attention will often involve the loss of considerable profit on the cow for the year. When purging is once fully seated from several days' neglect, it is often difficult to remove it.

The acidity on the stomach which always attends it must first be removed. A mild purgative medicine may be given. Rhubarb and magnesia is a very convenient article, and may easily be given in ounce doses along with the milk. Potash is also to be given in quarter-ounce doses in the same way. Two ounces of castor-oil, or two ounces of Epsom salts, might be given with the desired effect. After this, mild astringents may be given. Take prepared chalk two drachms, or magnesia one ounce, powdered opium ten grains, powdered catechu half a drachm, tincture of capsicum two drachms, essence of peppermint five drops. Mix together, and give twice a day in the milk or gruel.

After giving the above repeatedly without effect, which will rarely happen, take Dover's powders two scruples, starch or arrow-root powdered one ounce, cinnamon powder one drachm, and powdered kino half a drachm. Boil the starch or arrow-root in water till it thickens, and when cold stir in the other ingredients. Give night and morning. This complaint is often attended by inflammation of the bowels and general fever.

It is a good plan to keep a lump of chalk constantly before calves after they are two or three weeks old. It corrects acidity on the stomach, and is otherwise useful to them.

CONSTIPATION or COSTIVENESS sometimes attacks calves

a few days old, that have not been judiciously managed. It may be brought on by putting a calf to a cow whose milk is too old, or from feeding a calf from the milk of several cows mixed. It results from too heavy a mass of coägulated milk in the fourth stomach, which becomes very much swollen with hard curd. It is difficult to remedy. The best way is to pour down some Epsom salts, two ounces, dissolved in two quarts of warm water, by means of a horn or bottle, and follow this by half the dose every six hours.

Constipation sometimes appears in calves from two to four months old, when their food is too suddenly changed. The bowels must be opened and the hardened mass in the stomach softened very soon, or it will lead to fatal consequences.

Farmers are generally very careless about observing these things till it is too late. As already said, prevention is cheaper than cure ; but, if the complaint once appears, no time should be lost to administer a purge of salts in proportion to the size of the animal or the severity of the attack. Many a valuable animal will be saved by it.

The HOOVE often appears among calves after being turned out to pasture. The young animal coughs violently, and appears in pain. It should be removed at once to a dry place, and physicked. If taken in season, it is easily cured. If neglected, it will often prove fatal. This complaint assumes the form of an epidemic at times, and becomes very prevalent and troublesome.

Calves sometimes suffer from CANKER IN THE MOUTH, especially at the time of teething. The gums swell, and fever sets in. Common alum or borax, dissolved in water, may be applied, and a mild purgative administered, in the shape of one or two ounce doses of Epsom salts.

The diseases and complaints mentioned above are nearly all that afflict our dairy stock; and the list at least includes all the common diseases and their treatment. Some of the diseases and epidemics from which the cattle of Great Britain and other countries suffer are not known at all here, or are of so very rare occurrence as not to have attracted attention; and among these may be named pleuro-pneumonia, typhus fever, cow-pox, and various epidemics which have from time to time decimated the cattle of all Europe. To accidents of various kinds, to wounds, trouble with the eyes, and to lameness from other causes than those named, they are, indeed, more or less subject; but no work could anticipate or cover the treatment best in every case, and much must be left to the judgment of the owner.

I have tried to make this chapter, which I consider one of the most important of any to the dairy farmer, of practical value to every one who owns or has the care of a cow. But, lest a want of familiarity with some of the medicines recommended for particular diseases, or the fear of the expense of procuring and keeping them on hand, should deter some one from providing himself with a good medicine-chest, I wish to remind the reader that no small portion of them are always to be found in every well-regulated household, and that the others are obtained at so little expense that no one need be without them for a single day.

Let us see, for instance, how many of them are at hand. But few families are destitute of a supply of ginger, camphor, red pepper, lard, molasses, cinnamon, peppermint, starch, turpentine, tallow, bees-wax, burdock, and caraway-seed. The farmer's wife or daughter will generally have a supply of ammonia or hartshorn.

Now, I wish to suggest to the farmer or dairyman who happens to live at a distance from the apothecary

25*

to provide himself with a convenient little medicine chest, and put into it say four times the quantities of the various medicines which are mentioned in the preceding pages, carefully bottled and labelled for use. To aid in this simple plan, which might be the means of saving an animal worth twenty times its cost, I have obtained, from a *wholesale* druggist, about the average cost of the following quantities and kinds of medicines, which include all, or nearly all, that would be likely to be needed: Five pounds of Epsom salts, .18; one pint of castor-oil, .25; one pint of sweet spirits of nitre, .19; one pound of powdered nitrate of potash, .20; one pound carbonate of ammonia, .23; one half-pound sal ammoniac, .08; one pint of tincture of red pepper (hot drops), .37; one ounce of hydriodate of potash, .30; one pound chloride of lime, .10; one pound sulphate of iron, .10; 2 pounds powdered sulphur, .16; one pint of tincture of ginger, 37; one quart of essence of anise-seed, .50; one half-pound sulphuric ether, .20; one half-pound powdered sassafras-bark, .20; one quarter-pound magnesia, .06; one quarter-pound rhubarb, .30 (the common will answer instead of prepared); one ounce powdered opium, .43; one quarter-pound catechu, .06; one ounce Dover's powders, .25; 2 ounces gum kino, .05; one half-pound mercurial ointment, .37½; and one pound aloes, .25. Then keep in the chest a good probang, which is a flexible tube made for the purpose, and is much safer and better for introducing into the throat or gullet of an animal than a common whip-stick, which some use. This costs about $3.50, and can be procured at almost any veterinary surgeon's. This whole chest and contents will cost less than ten dollars.

Let the farmer also become familiar with the structure and anatomy of his animals. It will open a wide field of useful and interesting investigation.

CHAPTER XI.

THIS chapter I translate from an admirable little work in German, " *Die Holländische Rindviehzucht und Milchwirthschaft in Königreich Holland*," by Ellerbrock, a distinguished veterinary surgeon, professor of cattle pathology and cattle-breeding in the Agricultural Institute at Zeyst, in Holland.

MILKING AND TREATMENT OF MILK. — The cows are turned to pasture early in spring, and stay there day and night throughout the pasture-season. They are milked daily in a particular part of the lot called the milk-yard. This is kept in some instances permanently in the same place; in others, it is changed about at pleasure. A shady part of the pasture is generally selected, and it is commonly enclosed with a board fence. The cows are driven into this yard to be milked, when not already there at the usual time. The milking is done by male and female domestics, who carry their pails, cans, and dishes, hung on a kind of wooden yoke, Fig. 84, neatly cut out, painted, and set with copper nails.

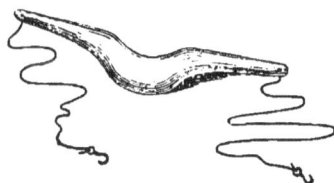

Fig. 84.

This is swung over the shoulders, or else the dairy utensils are carried on donkeys, ponies, or hand-carts ; or, where there is water communication, in boats, twice a day, to the yard.

In the larger dairies the utensils in common use are small wooden pails, Fig. 85, painted in variegated colors, with bright brazen or iron hoops, and neatly washed ; a strainer, Fig. 86, made of horse-hair; a large wooden

Fig. 85.   Fig. 86.   Fig. 87.

tunnel, Fig. 87, for pouring the milk into the cans and casks; one or more buckets, Fig. 88, usually of

Fig. 88.   Fig. 89.

brass, lined with tin, large enough to hold the milk of several cows together, or from twelve to eighteen quarts. In many dairies they have wooden buckets, Fig. 89, painted green or blue outside, with black stripes, and with iron or brass handles, kept very bright. Here the buckets are coated over inside with white oil-colors. These are borne by the yoke (Fig. 84), or in some of the ways indicated above.

In many places, instead of buckets for keeping the milk together, they use copper or brass cans lined inside with tin, and in the form of antique vases or large beer-jugs, Figs. 90 and 91, which are constantly kept brightly polished. In other places, they use for holding the milk smaller or larger barrels, Fig. 92, with broad hoops also kept constantly polished.

Instead of the yoke a soft cushion is also used, which
the dairymaids strap over their backs, so that they hang

Fig. 90.   Fig. 91.   Fig. 92.

down and rest over the hips and thighs. On this cushion the cans are laid, and fastened with broad hempen
straps, that they may not press too heavily upon the
body. This band is called the milk-strap. Where the
milk is carried home on a hand-cart, neatly-woven
baskets are fastened upon little wagons in which the
cans are placed. If it is to be carried in casks, the same
arrangement is fixed upon a hand-cart. Two wooden
floats are laid upon the milk in the buckets, in order to
protect it from slopping over. One or more large milk-
casks or tubs, in which it may cool off properly, are also
used. The size of these tubs is different, as well as the
materials of which they are made. Where the cooling
is not left to the air alone, but is sought to be effected
by hanging the milk-tub into cold water, the vessels are
made of metal. The large vase-like jars are also used
for this purpose. These hold about thirty cans, or
twenty-six quarts. Wooden bowls are used, of different
sizes and forms, and earthen pans, rather deeper than
broad, Figs. 93 and 94, in which the milk as it cools is

set for the cream to rise. A large pot for collecting the cream until there is enough to churn, and wooden skimmers for taking off the cream, are also used. The milker

Fig. 93

Fig. 94.

sits upon a common four-legged, and sometimes one-legged milking-stool, and milks either the teats on one side, or one hind and one front teat, the pail being held between the knees. The cows are milked regularly at four or five o'clock in the morning, and at five or six in the afternoon.

In West Friesland, North and South Holland, Utrecht, and other places, it is customary to tie the tail to the leg of the cow, that she may not annoy the milker. Most cows do not resist this, being accustomed to it from the beginning. They also pass a cord around the

Fig. 95

horns and tie her to a post stuck in the ground during the milking, as in Fig. 95. In many provinces only the unruly cows are tied in this way.

The milking takes place on the right side of the cow,

so that the milker sits on this side. In West Friesland
and North Holland there is an exception to this rule.
The cows are tied in pairs in the stalls, and one is
milked on one side and the other on the other, the
milker sitting with his back to the board partition, to
avoid annoyance from either animal.

When the milking is ended the milk is poured
through the hair strainer into the bucket, or through a
strainer or tunnel in the cans or casks, whichever are
used. The milk is taken to the dairy-house, without
delay, in some of the ways already mentioned. When
the yoke is used, one bucket is hung on the right side
and another on the left, each with a float on the top of
the milk to keep it from slopping over. The large
metallic milk-cans, with wooden stoppers, are borne
home on the cushions already described as being held
by shoulder-knots strapped round the waist. The
mode of transportation depends much on the distance
from the dairy-house and the quantity to be carried.

In winter, when the cows are in the barn, they are
likewise milked twice a day, and the milk is at once
strained through the hair strainer into casks made for
the purpose. These implements differ according to the
object pursued in the dairy; yet pans and pots are
mostly used for raising the cream to be made into
butter, since but few dairymen make cheese in winter.

All utensils necessary for milking, the preservation
of milk, and the making of butter and cheese, are kept
with the utmost neatness. Where a stream of running
water flows through the yard, the implements are generally washed in that, and flowing water is preferred for
the purpose. But where the farm or dairy-house
stands at a distance from a stream, a shallow fountain,
or basin, is dug out in the earth, walled up, and so
arranged that the water can be taken from it and fresh

water substituted when it gets impure. In such a
basin, or in flowing water, all new wooden dairy utensils are soaked for a long time before being used; but
those in daily use are washed, rinsed, and scoured out
with ashes, with the greatest care. None but cold,
clear, fresh fountain or flowing water is taken for cleansing dairy implements. It is to be observed that, in
large dairies, the use of water which is covered with
newly-fallen honey-dew, for washing the dairy utensils,
is carefully avoided. When the milk-vessels have been
perfectly rinsed out in fresh water, they are, in many
dairies, put into a large kettle of water over the fire,
and properly scalded; after which they are again cleanly
washed with cold water, so that not the least particle
of milk or impurity is to be seen, nor the least smell of
it to be observed. The metallic milk-vessels and the
metal parts of the wooden ones are cleansed with equal
care and exactness, and kept polished. Dairymaids
feel a pride in always having the brightest, most
polished, and cleanest utensils, and each strives earnestly
to excel the others in this respect.

When the milk-vessels are scoured, scalded, and
rinsed perfectly clean, they are hung on a stand of
laths and poles, made for the purpose, to be properly
dried. The round wooden milk-bowls, being made of
one piece, are very easily broken or split, and must be
handled with very great care in cleaning. To avoid
breaking, a peculiar table is used for scouring them.

The Dutch dairyman knows perfectly well that his
dairy can secure him the highest profit only when the
utmost cleanliness is the basis and groundwork of his
whole business; and so he keeps, with the most extraordinary carefulness, and even with anxiety, the greatest possible neatness in all parts of the dairy establishment.

DETERMINATION OF THE MILKING QUALITIES OF THE
COWS. — The Dutch cattle are, in general, renowned
for their dairy qualities; but especially so are the cows
of North Holland, which not only give a large quantity,
but also a very good quality, so that a yield of sixteen
to twenty-five cans * at every milking is not rare. Next
to these come the West Friesland and South Dutch
cows, from which from twenty to twenty-four cans of
milk may be calculated on. Though one could not
take a certain number and calculate surely what the
yield of each cow would be, yet he could come very
near the truth if he reckoned that a cow, in three hundred days, or as long as she is milked, gives, on an
average, daily, from six to eight cans of milk, from
which the whole annual yield would be from one
thousand eight hundred to two thousand four hundred
cans. Of this the cow gives one half in the first four
months, one third in the next three, and in the
remainder one sixth. These superficial results cannot
be taken, however, as the fixed rule.

Professor Wilkins, in his Handbook of Agriculture,
gives the following estimates of the yield of milk: A
good West Friesland or Gröningen cow will, after calving, give daily fourteen quarts of milk. This will, after
a while, be reduced to eight quarts. She may be milked
three hundred and twenty-three days in the year, and
her product in butter and cheese will amount to one
hundred güldens.

In Prof. Kop's Magazine it is stated that a medium-
sized Friesland cow, which had had several calves, was
giving daily, on good feed, five and a half to six buckets,
or from twenty to twenty-two cans, and over. In South
Holland, also, this quantity is considered a good yield

---

* A Dutch can is a little less than our wine quart.

of a cow. Of the cows of Gelderland, Overyssel, and
Utrecht, the yield cannot be reckoned higher than sixteen cans daily, and that only during the first half of
their milking season.

TREATMENT OF MILK FOR BUTTER. — To get good
butter it is quite necessary that the fresh milk be
properly cooled before it is set for cream. In the great
dairies of North and South Holland, which not only
possess the best cattle, but may be given as models in
dairy husbandry, they manage as follows:

The milk, as it is brought from the pasture, is poured
from the buckets, cans, and casks, through a hair
strainer, into one vessel, the milk-kettle. These milk-
kettles are not everywhere of the same size, or of similar form, but are always riveted together with strong
brass or copper bands, and lined with tin inside. The
most common milk-kettles hold sixteen cans; yet they
are found so large as to hold three barrels, or about six
hundred quarts. The peculiar kettle form is very rarely
found, but more frequently the cylindrical, or vase-
shaped. They are held either by two handles or one.
The number required depends on the number of cows
and the quantity of milk expected.

The milk-kettles, when filled, are set into a basin
with cold water, called the cool-bath, for the purpose of
cooling the milk. The cool-bath is frequently in the
kitchen, sometimes in the bauer-house, so called, or
directly before the cow-room, near the spring. The
latter is the most common and the most convenient
place. The water reservoir is dug in the ground, and
an oblong four-cornered form is preferred for it; the
sides of the excavation being walled up with hard-burnt
building-stones and cement, but the bottom is laid in
tiles, either red, hard-burnt, or white glazed. Richer
dairymen take finely-hewn blue stone or white marble

for it. The size of the reservoir is governed by the number of milk-kettles to be put into it, and so is its depth by their height, so that the rim of the kettle is on a level with the top of the cool-bath, Fig. 96. The sides of the cool-bath in the kitchen project some feet over the floor, yet are not so high that the setting in and taking out the milk-kettle will be at-

Fig 96. Cool-bath.

tended with great inconvenience and trouble. Where it is desired to make the work of setting in or raising up the milk-kettles from the cool-bath as easy as possible, a beam is fixed along the side of the trough, and iron props are firmly fixed, which extend out a little over the edge of the trough, half-way down from the beam. On these the operator can support himself in lowering or raising heavy vessels. These stays, or props, are sometimes fixed directly into the wall, along which the cool-bath stands. Under the bottom of the reservoir, on the other side from where the water comes in, is an outlet, stopped with a tap or faucet, to let off the water.

The cool-baths in the kitchen are, for the most part, on the floor, and extend up a convenient height; whilst those in the cow-barns, as a general rule, are dug down and walled up, and their top is fastened to the floor of the barn. They are deep enough to allow the water for cooling the milk to come up to the rim of the milk-

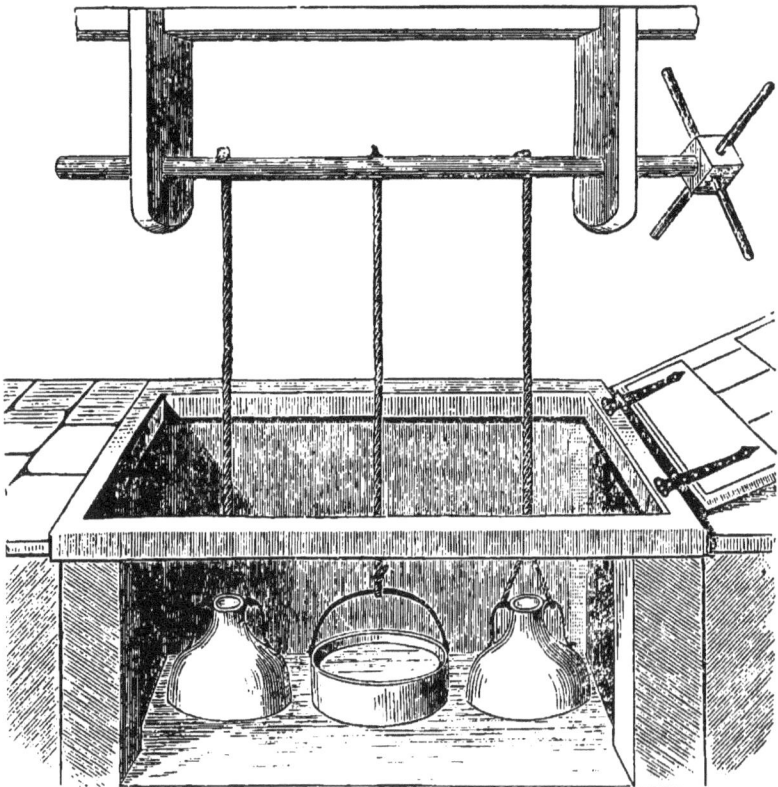

Fig. 97. Cool-bath

kettle; but, in order to prevent men and cattle from falling in, it is covered with a strong wooden lid to shut down, as in Fig. 97.

Such a cool-bath is used in the cow-room only in summer, when the heat is so great that it is difficult to keep the milk cool in the kitchen. The cool-bath in the cow-room is considered as only an auxiliary to that in the kitchen, and to be used only in case of necessity. The milk-kettles are hung by their handles, and let down by means of a crank. When the platform is not in use it is taken away from the cool-bath, and the cover is let down and kept closed.

The milk is allowed to remain in the cool-bath until the froth has disappeared, and there is no difference in temperature between the water and the milk. The milk of one milking must give place for the next, so that it will be changed twice daily, morning and evening. A very great importance is, everywhere in the Dutch dairies, attached to this rapid cooling of the milk, because it is known by experience that it is thus greatly protected from turning sour.*

The milk, when properly cooled, is brought to the milk-cellar, where it is immediately poured out of the milk-kettles into vessels designed to receive it. Wooden bowls or pans, or high earthen pots, are used for holding it. The pans and pots are set on the table, and a small ladder, or hand-barrow, is laid on them, on which is placed the strainer, when the milk is poured from the kettles. The wooden milk-pans are of several forms, generally made of ash or of linden, and oval. They are, on an average, three and a half feet long, and half a foot broad, more or less; but their dimensions vary.

---

* It will be perceived that the arrangement for cooling the milk before setting in the pans, in the Dutch dairies, is very elaborate. I have followed the original in translating the above, though the practice in Holland differs widely from our own in this respect, and from that recommended in the preceding pages. The point may be worthy of careful experiment. — TRANSLATOR.

It has been found, by experience, that the flatter and shallower the pans, the quicker and better the cream rises. The milk-pots are pretty large, but are rather shallow than deep, glazed inside, of different forms, and different capacities; but they are always broader on the top than at the bottom, though they stand firmly on a round, broad foot-piece. Milk pans and pots are rinsed with cold water before the milk is poured into them. When properly cleaned and filled, they are placed on shelves made for the purpose, in regular rows. These shelves are only a few feet high above the floor of the cellar, and of suitable width; but, if there is not space enough for the milk, the pans are placed on the bottom of the cellar. The pots are also set along the walls, on firm board shelves.

The milk-cellar, or rather the milk-room, Fig. 98, in the North and South Dutch dairies, is placed on the north side of the house, next to the kitchen, but a little lower than the latter, so that there are usually three steps down. The longer side, facing towards the north, has one window, whilst the gable end, with its two windows, faces towards the west. The windows are generally kept shut, and are open only nights in summer. The cellar is either arched or covered with strongly-boarded rafters, over which the so-called cellar-chamber is situated. The floor of this room is laid in lime or cement, with red or blue burnt tiles, so that nothing can pass down through into the milk-cellar. In the cellar itself are the above-mentioned shelves and platforms for the milk-vessels along the walls, while outside, in front of the cellar, linden and juniper trees are planted, to prevent as much as possible the heat of the sun from striking upon the walls. Cleanliness, the fundamental principle of Dutch dairy husbandry, is carried to its utmost extent in the cellar. Barrels of

Fig. 98. Dutch dairy-room.

meat, bacon, vegetables of every kind, and everything which could possibly create a strong odor and infect the air, or impart a flavor to the milk, butter, or cheese, are carefully excluded.

The vessels in which the milk is set remain standing undisturbed in their places, that the formation of cream may go on without interruption. Twenty-four hours, on an average, are thought to be necessary for the milk to stand, during which time the cream is twice taken off, once at the end of each twelve hours. The morn-ing's milk is skimmed in the evening, and the evening's on the next morning. But the milk always remains quite still till the dairymaid thinks it time to skim, which she decides by the taste. Long practice enables her to judge with great certainty by this mode of trial.

Fig. 99.

When the cream is ripe it is taken off by the dairymaid with a shallow wooden skimmer, Fig. 99, in the form of a deep plate, and carefully placed in a particular vessel — a bucket or cream-pot. The cream-pot is generally washed very clean, the staves very finely polished, striped with blue or white outside, and held together by broad brass or copper hoops, kept very bright. For closing the jar they use an ashen cover, which is either simply laid on by a common handle, or sometimes held on by brass or copper hinges. Both cream-pot and cover are always scoured quite white and clean. The cream remains there till enough is got for churning, or till it becomes of itself thick enough for butter. It is known to be of the proper consistence for butter when a long, slender, wooden spoon, thrust down into it, will stand erect. When in summer the cream does not get thick enough in season, they seek to hasten it by putting in a little butter-milk; but in winter the ripening of the cream is hastened by warming, either by holding the cream-pot over a coal-pan, or on a hearth-plate.

The remainder, the skim-milk from the milk bowls or pans, sour milk, or butter-milk, is poured into a particular vessel, and made into spice-cheese.

Besides the methods here described for keeping milk for butter, milk is used for other purposes. Sweet milk cheese is made of the unskimmed milk: cream is used in the house for coffee. Rennet is also added to fresh milk, and the product is immediately sold, being greatly relished by many. From skim-milk and butter-milk put together is made an article called kramery, by cooking the mixture, putting it into a linen bag, and hanging it in a cool part of the milk-cellar, or elsewhere, when the liquid drops out and leaves a mass of considerable consistence, called Hangebast.

As soon as the milk is taken from the vessels, they are taken out of the cellar and carefully cleansed and dried before being used again.

METHODS OF CHURNING. — Churning is the principal operation in the manufacture of butter, for by it the fatty particles are separated from the other constituents. There are several methods in Holland of effecting this separation of the butter globules. The oldest and simplest is that of putting the cream into an upright churn, in which the cream is agitated by moving a long dasher, pierced with holes, up and down, till the object is accomplished.

There are, strictly speaking, only two forms of the churn which are used in all parts of the country. One is broad at the bottom and narrow at the top. This has been known from the earliest times, and is called the old churn, Fig. 100.

This old churn is still used in many dairies, and it

Fig. 100.

has the preference over the other form, because it is thought to bring the butter quicker and more completely.

The other form is more like a beer or brandy cask on end, being smaller at each end than in the middle, and is called the barrel-churn. Both kinds are made of oak-wood, and have wooden or broad metal hoops. In the one case they are painted outside ; in the other, they remain of the natural color, but are the more frequently scoured, so that the dark-colored oak-wood gets a whitish color. The metallic hoops are always kept polished bright.

Both kinds are of different sizes, according as the quantity of cream is greater or less, or as they are to be worked by hand or animal power simply, or by machinery. In South Holland, where unquestionably the most butter is made, the barrel-churn is at each end about two feet and two inches in diameter, and in the centre is seven inches broader, with two-inch staves. The old churn, on the other hand, is usually fourteen inches at the top and twenty-five at the bottom.

In North Holland and West Friesland, also, sizes are found in which one hundred and fifty to two hundred quarts of cream can be churned. The churns have each a strong cover at the top, which fits into their rim about the thickness of the 'hand, with a hole in the middle for the dasher.

The churning is performed either by the hand motion of the dasher, as in all small dairies, and in the smallest churns, or by man-power with the help of certain mechanical contrivances. The means for effecting this are different, and so the churns have different names.

In many dairies, for instance, they have a lever connected with the dasher; in other places they use a
flexible pole, fixed into the ceiling above, for facilitat-
ing the motion of the dasher, or put a lever in motion
with the feet, which raises and sinks the dasher. There
are also complicated artificial butter-machines and
butter-mills, which are named after the inventor, the
manufacturer, or the motive power. The most known
and widely used are the turning-mills, the wheel-mills,
and the clock-work mills; as the Hand Butter-Mill of
Valk, Fürst's churn, etc.

There are also still more elaborate machine-works
for moving the dasher, which are used in the larger
dairies on account of their convenience and economy.
Dog-power and horse-power churns are frequently met
with.

CHURNING IN THE COMMON CHURN. — The use of this
is well known. The dasher is moved up and down by
hand, with the churn full of cream, till the butter
particles are separated and collected together. The
operator keeps his body in equilibrium, to exercise the
power of moving the dasher regularly for agitating the
cream.

THE LEVER CHURN is very commonly used in South
Holland, Fig. 101. The churn itself is barrel-form,
as already described, and the dasher is put in motion
by a lever. The upper end is pierced with holes,
through which runs an iron pin. In a beam of the
ceiling two joists are firmly fixed, about a foot and
five inches long and four inches square, and several
inches apart. The longer arm of the lever is four
feet and seven inches; the shorter, three feet and six
inches. The churn stands under the short arm of the
lever, where the dasher is fixed. By drawing the
longer arm of the lever towards him, the operator

Fig. 101.

presses the dasher down through the cream.   This
mode is far less wearisome than the hand-churn,
because by the lever, with less expense of power, a far
greater agitation is produced.   A weight is sometimes
attached to the longer arm, by which the power required
is still further reduced.

CHURNING WITH AN ELASTIC ROD. — The old-fashioned
churn is set in motion by the aid of another kind of
power, as seen in Fig. 102.   A long, tough, flexible
stick is fastened into the cross-beam in the ceiling, so
that its larger end is held firm by two iron clasps.   The
elasticity of the rod is such that, when the smaller end
is drawn down by hand, which, at the same time, moves

the dasher, it rebounds, and thus saves considerable expenditure of power.

Fig. 102.

CHURNING WITH THE TREADLE LEVER. — In many places the churn is put in motion by the feet, as in Fig. 103, where several levers are united to produce the upward and downward motion of the dasher. The longer arm of the lever is connected with the churn, and the shorter is set in motion by a foot-board. The foot-board lies on a roller, with its longer part attached to the lever; and by throwing the weight of the body upon this part the shorter arm of the lever is drawn down, and the longer, attached to the churn-

27

Fig. 108.

dasher, is raised. The mode of operation is so plainly
seen in the cut as to need no explanation.

Among the more ingenious contrivances used for
churning in Holland belongs the churn invented by
Fürst. The body is somewhat similar to the barrel-
churn, but is smaller; and it is of uniform diameter
throughout, as in Fig. 104. It is covered with a
wooden lid, furnished with a convenient handle, and
stands on a low platform, to which it is fixed, when in
use, by means of a screw, *k*. The motion is communicated to the dasher by means of a wheel, or windlass, and an endless cord.

In the interior of the cylinder is placed a kind of

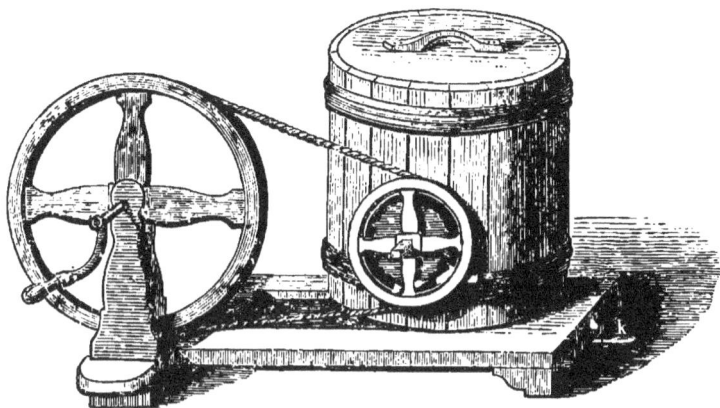

Fig. 104.

ventilator, Fig. 105. This consists of eight wooden
wings, pierced with holes, and motion is communicated
to it by means of the wheel, *b*, connected by the

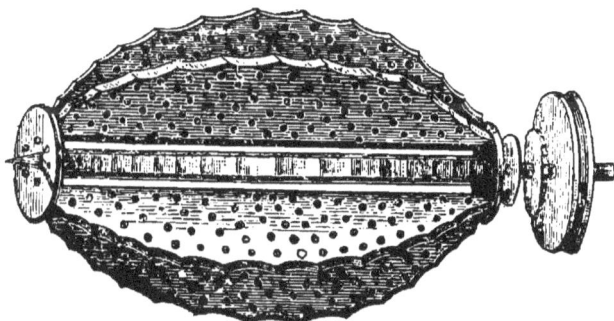

Fig. 105.

cord to the larger windlass. The wings of the machine,
when set in motion, strike incessantly in the cream, and
so powerfully that the whole mass is agitated, and in this
manner the separation of the butter particles is soon
effected. The motion is so rapid that it is often necessary to turn the crank very slowly, especially just as the
butter is coming.

VALK'S HAND BUTTER-MILL, Fig. 106, has many advantages. It is less fatiguing to work than the old-

fashioned churn, and even than Fürst's, because the
motion of the body required is simple and less exact-
ing.   And again, the churn takes up less room; and is

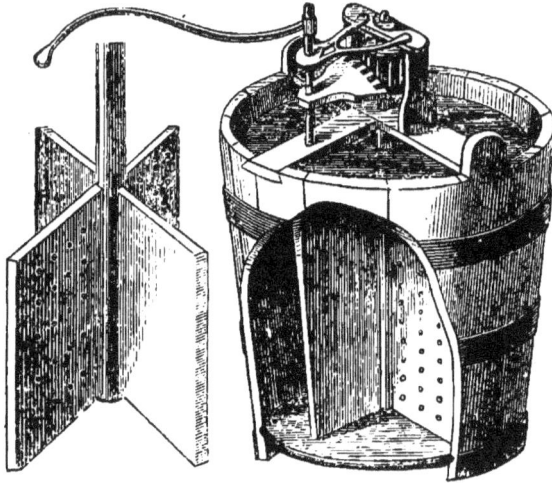

Fig. 106

easily transported, which is an important consideration
in churning, on account of the influence of the tempe-

Fig. 107.

rature. In summer the heat may delay, or render the operation difficult, and in winter the coldness presents obstacles. A transportable churn can be moved into a cool place in summer, and a warm one in winter, when it is desirable. The dasher of the churn is also seen separate in the same figure.

THE DOG-POWER CHURN, Fig. 107, economizes labor, while, at the same time, more butter is obtained, on account of the uniformity of the agitation produced. It is in use in all the Dutch provinces. The form and size of the churn are comparatively indifferent; but the tread-wheel and direction of the moving power are the important points. The diameter of the wheel is from ten to twelve feet, and the rim or outer circumference is made of boards two feet wide. The weight of the animal turns the wheel and moves the dasher by means of cogs, as shown in the figure.

Where there is a sufficient supply of moving power, a churn with two dashers is sometimes attached, as shown in Fig. 108, in which case one dasher moves down while the other is raised.

A large and strong dog is required, and he is easily taught to keep to his work, by beginning with short trials, and gradually lengthening them. A steady and uniform step

Fig. 108.

27*

is necessary, and this will soon be acquired.   The dog
is sometimes left free, and sometimes tied by a line.

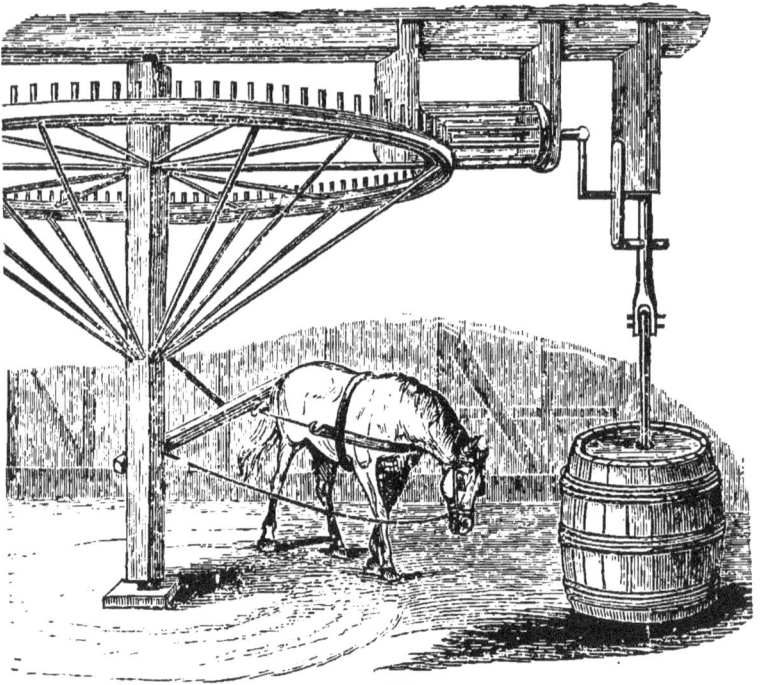

Fig. 109.

CHURNING BY HORSE-POWER. — On large farms and in
extensive dairies the churning is done by horse-power,

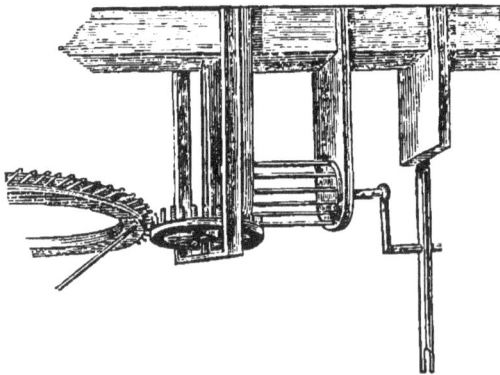

Fig. 110.

as shown in Fig. 109. The form of the churn itself is optional in this case, also. The size of the wheel varies, but it is seldom less than nine or ten feet in diameter, furnished with cogs on the upper surface, which are from four to six inches long, and play into a smaller wheel, the axle of which is attached to the dasher of the churn. A third and smaller wheel is sometimes introduced, as in Fig. 110. A quick and regular step is required of the animal, and a quiet and docile horse is always preferred. A horse adapted to this work commands a good price. Blinders are always used on the horse while churning.

DURATION OF THE CHURNING. — In whatever way the churning is performed, the result is always a separation of the fatty particles from the other constituents of the milk. As soon as the churning indicates that the butter particles increase in size and collect together, the motion of the dasher must be hastened till the butter has come together in a large mass. Great care should be taken to observe the appearance of this formation. The Dutch dairymaids acquire great skill, by long practice and experience, in judging of the proper moment when the separation of the particles has completely taken place. Very great importance is with justice attached to this skill, for it is undoubtedly true that one with this knowledge can get far more and better butter from milk of the same quality, the same quantity, and skimmed at the same time.

The cream taken from the milk of thirty-five cows, after standing twenty-four hours, is generally churned in summer in less than an hour, sometimes in three quarters of an hour. In very hot weather the cream-pot is frequently set into the cool-bath of fresh water for five or six hours before the churning begins, and it churns the easier for it. Cold water is never poured into the churn with the cream. In winter, as well as in cold weather in spring and fall, warm water is some-times poured in with the cream.

WORKING AND TREATMENT OF BUTTER. — When the churning is finished, the dairy-woman takes out the butter with a wooden scoop, Fig. 111, and puts it into a tub for further working. The tub.

Fig. 111.

Fig. 112, is a broad, shallow vessel, open at the top, and having an opening at the bottom which is stopped by a bung. The scoop is pierced with holes, through which the butter-milk

Fig. 112.

drains. The butter put into the tub is now rinsed. salted, and formed.

The tub is put upon a low, firm table, and the butter is worked by the hands, or by a shallow, rather wide and strong wooden ladle, until the butter is united into one firm and entire mass. Many dairy-women are accustomed to work the butter out from the mid-dle towards all sides before bring-ing the whole mass together in the tub. Then very clear and pure fresh cold water is poured upon the butter, and worked through it till all the milky particles are entirely removed. After this is done in several workings, the bung is removed from the bottom of the tub, and the watery

Fig. 113.

matter runs down through a little strainer, as in Fig. 113.

As a general rule, butter is washed with water and worked over eleven or twelve times; yet the operator must judge whether the butter contains any particles of milk, and must work with water till, as it runs off, it is no longer whitish, but perfectly clear. Butter sometimes becomes too soft from too much working, if it is all done at once; it is then worked over two or three times, and allowed to stand in cold water after each working, which preserves its hardness and texture. This whole operation is called the washing of the butter.

When the washing is finished, the butter is cut with a blunt, saw-toothed knife, Fig. 114, in every direction, in order to remove all hairs, or fibres of any kind, which by any possibility have got into it during the day. It is then sprinkled over with white, finely-powdered salt, the quantity of which is regulated by the taste; and this is perfectly worked in, so that the whole is uniformly salted. Most dairy-women determine the quantity of salt by the eye and the taste, and acquire such facility by continued practice that they always get the proper quantity; but less experienced ones take the salt by weight. The salting is not all done at once, but is continued three or four days, twelve hours intervening between each application, until all the salt has dissolved, and not a crystal is to be found. If the butter has a speckled and variegated appearance, it is a sign that the salt is not completely worked in, and the neglect must be remedied by working it over still more in the most thorough manner. When the salt is all dissolved, the butter is brought into single balls and got ready for the next market-day, or the whole mass is put into a particular keg, in order to be taken to market at some subsequent time as firkin-butter.

Fig. 114.

The Form of Fresh Butter. — The form of the butter is made by taking a suitable quantity and pressing it into a mould, and then taking it out by knocking
on the mould. Many different forms of butter-moulds
are in use in the different sections of Holland, such as
are shown in Figs. 115, 116, and others.

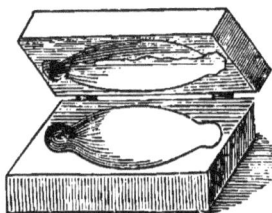

Fig. 115.   Fig. 116.

The figures impressed on the butter are given by the
mould, where it is deeply engraved ; or they are made
after the butter is taken out of the mould, and for this purpose a peculiar instrument is used, Fig. 117, a kind of flat
wooden spoon, with a short, convenient handle,
and long grooves in the broad, flat surface. Each
region has its own peculiar stamp, or special
figures, which are given to lump-butter, to which
particular attention is paid by the purchaser.
The butter-dealer knows exactly that in one
section butter is stamped in one way, in another
section in some other way ; and that the butter
of one section, with its peculiar stamp, is worth
more than that of another.

Fig. 117.

The butter-moulds are generally made of linden-wood,
but must always be large enough to hold at least a certain prescribed weight of butter; for all lump-butter
brought for sale to the weekly market must be of a
prescribed weight. This weight is very different, and
almost every city has different regulations and market
customs; yet, in most places, a pound is the legal

weight. Certain market-masters, or inspectors of butter, are appointed, and watch that all the butter has its proper weight. If too light, it is forfeited by the seller, who is also punished for fraud. The butter brought to market is generally covered with very clean white cloths, and several sample lumps are put for inspection in a large butter-bowl, basket, or shallow box.

Many dairymen are accustomed in spring, when the first grass butter is made, to send their regular customers a few little lumps of fresh May or grass butter. These presents generally have a peculiar form, and on the specimens most carefully prepared some animal is moulded, as a sheep lying down, a dog, &c., with a bunch of green grass or buttercups in its mouth. The dairywoman herself usually presents this butter in a beautiful milk-bowl adorned with grass and flowers, covered with glittering white cloths.

THE PACKING OF BUTTER IN FIRKINS AND BARRELS.— If the butter packed in firkins and barrels is to be kept a long time, experience and knowledge are required to pack it so that it will not be injured. The form and size of these casks are different in different sections and provinces. Where butter-making forms a chief branch of dairy business, the particular form and size which have been used for a long time are adhered to, because, dairymen know very well that the public recognizes' their choice butter by the form and size of the casks, and buys it the more readily. The greatest anxiety of the Dutch butter-maker is to keep up the old, well-earned reputation which Dutch butter has in every foreign country, both for its intrinsic good qualities, the result of the process of manufacture, and for its extraordinary appearance as an article of commerce.

For the proper preservation of the good qualities of

butter, it is of the highest importance to have the casks properly made and treated; but the mode of salting and packing the butter in them is also of special importance, since this is examined at the sale. The old and customary forms and sizes of butter-casks are, therefore, of great consequence to the butter-maker, because every butter-dealer and judge of butter recognizes at once, by the external form of the casks, from what section the butter comes, and makes up his mind on the money value of the article from these appearances.

It was not originally known what kinds of wood were best for transporting butter long distances in, and preserving its highest qualities; and butter-casks were made of several kinds of wood, as oak, beech, willow, etc. But it was for the interest of the government that Dutch butter should maintain its reputation for extraordinary qualities abroad, and the most rigid laws were enacted, prescribing from what wood the casks should be made, etc.; and now only oak is allowed to be used, and the casks are all inspected and stamped according to law. * * * *

Before the butter is packed the casks are properly cleaned and prepared, for which practice and experience are requisite.

Old butter-casks that have been previously used are cleaned of every particle of fat and dirt remaining in them, and scoured and washed out as carefully as possible, and are placed for several days in running water before they are used again. If no running water is at hand, quite clean pond or spring water is taken, and all impure water is carefully avoided. After they have lain in the water five or six days, they are carefully scoured out with good wood-ashes and sand, and again well rinsed. After several scourings and soakings, they are put into a kettle over a fire and carefully scalded; and then, when cold, again scoured and rinsed, for which the most judicious dairymen use milk instead of water, and they are then placed to dry in the air. They are fit for use only when everything has been done in the most careful manner.

But new butter-casks require still more particular and careful treatment before they can be filled with butter without fear of injury. They are got ready for packing in several different ways. Some dairymen let them lie in pure water a whole summer and winter long, and wash them out in lye, and then treat them just as they do those that have been used. Others, however, who give the new casks the preference over the old, but who cannot wait for the soaking in lye over summer and winter, treat them in the following manner: They prepare a lye of good American potash, which generally contains the most alkali, in a cask holding some three hundred quarts, taking a pound of potash to twenty pounds of water. For a cask of the size named fifteen pounds of potash are used, which is prepared by pouring boiling water upon it and stirring constantly, adding a little more water as the potash dissolves. With this lye, which will be about five degrees strong by Beaumé's aërometer, the butter-barrels are entirely filled. The barrels stand two hours filled with lye, and are then emptied and exposed to the air to dry, without being scoured out with water or milk. The lye may be used again for other new barrels, even though a part of its strength may be gone. Potash is added, from time to time, to keep up the specified degree of strength. A solution of fifteen pounds of coarsely-powdered alum is prepared in about three hundred quarts of hot water, in a vessel as large as the lye-cask. The butter-barrels are also filled full of the solution of alum, and allowed to stand twenty-four hours. This alum solution must

28

also be of five degrees strength by Beaumé's scale, and
it can be used over and over by adding more alum now
and then.   After emptying out the alum and lye, they
are dried a day in the sun and air, and then rinsed out
in fresh, pure water, when they can be used for packing
butter without fear.   Some add a little sulphate of
iron or green copperas to the alum, when the solution
is more powerful ; yet the management of the butter·
barrels is then more troublesome, and requires more
experience.   The effect of the copperas has also the
disadvantage that it blackens the barrels, which, though
it does not injure them, is not liked by the purchaser.

By this treatment the new butter-barrels are much
more quickly and cheaply cleansed, and got ready for
packing and transporting butter, than by the course
pursued with old barrels.   The barrels, treated as above,
are not only quite water-tight, but the wood is stronger
and more durable.   By means of the potash-lye and the
alum solution the tannin is taken from the oak-wood
used in the barrels, which, if it remained, would give a
disagreeable taste to the butter.   The effect of the potash and alum upon the wood of the barrels is quite
harmless, and does not impart the least unhealthy quality
to the butter.

When the old or new barrels have been cleansed and
prepared, in either of the ways indicated, suitably for
packing the butter, the bottom of the barrel is evenly
covered with salt.   Then a layer of butter which has
been thoroughly washed and salted is made, and
another layer of salt, and so alternate layers of salt
and butter till the barrel is full, when a little brine of
salt and water is poured on top.   The butter is now
ready to be laid in the cellar, and thence to be sold and
exported.   When the dairy is not sufficiently large to
fill a barrel each day, the butter of several churn-

ings must be used, and the barrel filled from time to time as it stands in the cellar.   In that case the upper layer of butter is left covered with salt, and the cover of the barrel is closed down tight.   In most large dairies a barrel is generally filled at one churning, which is considered better for the quality of the butter.   The butter is always packed in so firmly that no space is left unfilled.

In doing up butter for sale at home, or at a neighboring market, the lumps are worked into the form of half a sphere, and put into little bright-hooped boxes, made to fit into larger casks, which can be nicely covered and closed up, as seen in Fig. 119, where the dairy-woman holds a box in her hand.   The covered casks are also seen carefully nailed up.

Fig. 119

The buyer who wishes to try the butter uses a long iron or steel borer, hollow inside, and furnished with a handle, as also seen in the cut.   This not only enables him to test the quality but the uniformity of the butter in the cask.

COLORING OF BUTTER. — The practice of coloring butter is founded on. the fact that we are accustomed to
form our judgment at once of the qualities of the article from the whiteness or the yellowness of its color.
Whiter butter is less attractive generally than yellow
summer or grass-made butter.   The color has come to
be important to the seller, and artificial means are found
to regulate it.

The coloring is made as follows: About a pound of
butter is melted, so that the heavier parts sink to the
bottom, when the light, clear fat on the top is poured
into another dish.   In this fat thus poured off is put a
piece of annatto about the size of a walnut, wrapped
up in a linen cloth, and it is then again put over the
fire. The coloring matter of the annatto strains through
the linen cloth, and turns the butter brown red, when
it is allowed to cool off.   When the butter is to be colored, some of this brown red is melted, salted, and
mixed very carefully into the butter after washing. The
quantity of coloring matter used depends on the color
which the maker wants to impart to his butter, and a
little practice soon enables him to take the right quantity.   Others pour the coloring matter directly upon
the butter to attain the same end.

In coloring artificially it is important to get a uniformity of color, which is the result of very thorough working.   Colored butter must not be marbled.

The cream is sometimes colored before churning.
The annatto is put into a clean beech-wood lye, and as
much of this colored and strained lye is taken as is
necessary to produce the desired color in the butter. It
is then churned as usual.

Turmeric is sometimes used instead of annatto for
coloring butter. It has no advantage, however over
annatto

In many sections the butter is colored with an extract of saffron in water, or of marigold, or with the juice of carrots, which is applied to the cream before churning.

The coloring adds nothing to the quality or the taste. It is done for the sake of the looks; but it gives the butter a deceptive appearance.

USE OF THE BUTTER-MILK. — The butter-milk in the churn is poured into a great cask, which in large dairies, as a general rule, is painted blue outside and white inside, with broad black iron hoops. It stands generally in the kitchen covered with a wooden lid. Butter-milk is used either in cooking, or for calves or swine, or is sold.

Dairymen in the vicinity of large cities have barrels with broad, bright brass hoops, in which they carry their butter-milk to market. It is put into them through a bung-hole, and when full the wooden bung is wound with linen and driven in. In these barrels the butter-milk is carried to Amsterdam, Rotterdam, etc., sometimes by boats on the canals, sometimes on wagons, and by yokes, and there sold to the grocers at wholesale, to be again sold out by them. The butter-milk thus brings an income by no means inconsiderable to well-managed dairies.

THE MANUFACTURE OF THE DIFFERENT KINDS OF DUTCH CHEESE. — From time immemorial, cheese, as an article of commerce, which has had a large sale, has brought an extensive income to the cattle-breeders and dairymen where its manufacture has been largely carried on, as everywhere in West Friesland, North and South Holland, and along the borders of the crooked Rhine in Utrecht.

Dairymen are not the only ones who enjoy the advantage which grows out of the cheese-trade; but a large

28*

number of other people derive considerable profit from
it, and support themselves entirely by it. Even the
commonalty of the cities, where the weekly markets for
the sale of cheese are regularly held, derive a considerable revenue from the small taxes for carriage and
market-dues, to which every seller has to submit.

The actual difference between the different kinds
of cheese made in Holland is due in part to the form
and size, and in part to the mode of making. Every sort
has also a name derived from its peculiarities, or from
the provinces or sections where it is made. The varieties of cheese best known in the markets in South
Holland are the spice cheese, the sweet milk cheese,
known also under the name of Gouda cheese, the so-
called May cheese, the Council's cheese, the Jews'
cheese, and the English cheese, made in many places.

Further up in North Holland, the North Dutch sweet
milk cheese, as it is commonly called in the province,
known in the foreign markets as Edam cheese, is almost
exclusively made. A kind of sweet milk cheese is made
to a limited extent, called Commissions' cheese. In
West Friesland, Utrecht, and South Holland, but few
except sweet milk cheeses are made.

In making cheese, the utmost cleanliness is most carefully observed in all the operations. Whoever is
intrusted with this work is required to display the
utmost neatness in his whole person, as well as in the
dairy-room; and the vats and other utensils are daily
scoured, washed with lye, and washed out in water and
rinsed. The greatest attention is also paid to the trans-
port of cheese to the weekly markets in the cities;
and in whatever way his load is carried, whether by
wagon or in little boats, the person intrusted with it is
always dressed in the so-called cheese-frock, a large white
linen, which is used exclusively for this purpose. At

the market itself the cheese is laid on a four-cornered bench, two feet high, and exposed to view in a glittering white linen cloth. But, in order to keep off all dust and impurities, a sail-cloth is raised over the whole, called the cheese-sail; or it is covered with a sail-cloth covering, or sometimes with clean straw. But in other places it is customary to carry the cheese on wagons, in a white linen cloth, and covered with a woollen cover, ready packed for sale at the markets.

CHEESE-MAKING IN SOUTH HOLLAND. — Spice cheese from skim-milk, and sweet milk or Gouda cheese, are the only kinds made to any extent in South Holland. Spice cheese, which derived its name from the addition of spices, is a firm, flat cheese, of about twenty pounds weight, brought to market generally colored red. It is three quarters of a foot thick, and one and a half feet in diameter, and is made as follows:

The skim-milk is poured from the milk-pans into large tubs, and allowed to stand quiet till the cheesy matter has settled to the bottom, which requires, perhaps, half a day. Then the thin liquid on top is poured off very carefully, without stirring up the rest, through a strainer, into a large brass kettle, till it is full; but the thicker substance at the bottom is left, and not put into the kettle. Under this kettle a fire is made, and the milk heated to a certain degree, regulated by the judgment of the dairymaid, sufficient to warm other cold milk, but it must not boil. The fire is made in the kitchen, or in the summer-house, or in some other room called the cheese-house. When the milk in the kettle is properly heated, it is poured into the tub of milk which has been heated and allowed to get cold. This tub is an upright vat, open at the top, of uniform diameter, bound with wooden hoops, and generally left of the natural color of the wood; scoured very bright, but some-

Fig. 120.

times painted blue and the hoops black. It is seen in Fig. 120.

When the quantity of milk is large, the dairyman puts in as much rennet as he thinks necessary to curdle the milk completely; but before and during the addition of the curd the whole is thoroughly stirred, and this stirring is continued until the stick or wooden ladle used for the purpose will stand erect in the curd. Then the dairywoman works the curd with her hands till no further effect of the rennet in curding the milk is to be seen. It is called the cheese-curd.

The rennet is prepared in the following manner: The maw or fourth stomach of a newly-killed sucking calf is taken from the other stomachs, carefully cleaned and cut into strips two inches wide, and then hung up in the chimney to be smoked and dried; or, in hot weather in summer, it is hung up in the sun. Well smoked and dried strips will keep a very long time. When these are wanted for use, they are very carefully washed and purified, and then laid in the salt brine from the butter-barrels, or in lukewarm salt water to soak. The liquid is put into bottles and laid in the cellar. For curding milk as much is taken as is thought to be necessary, which cannot be determined without considerable practice and experience. If too little is taken, the cheese is not fat enough; if more than the right quantity, it gives a disgusting acid taste. It is difficult, almost impossible, to state exactly how much rennet should be used with a certain quantity of milk, because this must be determined by its quality and its strength. Something like the following quantity is, however, taken : In a sixty-quart vat are placed about fifty rennets, prepared by drying, washing, and cutting, and a clear salt brine or butter-pickle of twenty to twenty-five degrees strength is added. In smaller quan-tities the proportion of rennet is about one and a half quarts to a rennet, or even less. This dried maw can be bought everywhere in packages of twenty-five pieces each.

One great point in cheese-making is to have a suffi-cient quantity of good rennet in store ; for the older it grows the more powerful and effective it becomes, and the experienced cheese-makers, studying their own interests, know very well how difficult, hurtful, and time-wasting, it is to use fresh or new rennet. The asser-tion sometimes made that they use muriatic acid instead of rennet for curding the milk in Holland rests on an error, at least so far as the present methods are con-cerned. In earlier times, and for the poorest kinds, as the Jews' cheese, muriatic acid was more or less used.

At the present time, the rennet for those cheeses is prepared from the stomachs of calves some days old.

When the curd has sufficiently come, and has all been thorough-ly broken, the dairy-woman puts a four-cornered linen cloth, called the cheese-cloth, which is used only for this purpose, and is only loosely woven, upon a small strong ladder laid

Fig. 121.

over the edges of a low tub, and puts upon the cloth the proper quantity of curd, then ties up the four corners of the cloth, and presses with her whole strength, that the milk may drain off. This work is also done by men who can apply great strength, Fig. 121. The corners of the cheese-cloth are brought together, and the operator presses as hard as he can, in order to remove all the milk from the curd. But, as this is not possible with the hands alone, the whole is placed under a plank-press, and by this means as much of the milk as possible is pressed out. A strong cleat is nailed to a pillar in the wall at a convenient height from the floor,—say two feet,—so that the tub, ladder, and cheese-cloth, can be put under the plank, when the plank is pressed down upon the cloth and curd. At the other end of the plank the operator sits and presses

Fig. 122.

down with the whole weight of his body, as seen in Fig. 122. The whey runs into the tub, and is generally used as food for swine. The pressure is continued till no more runs off.

After the complete removal of the whey, the curd remaining in the cloth has the form of the palms of the hands, and is pressed so firmly that it holds together when the cloth is removed. But it is again broken up, and put for this purpose into the breaking-tub, a low but broad, open tub, with wooden hoops, and made of strong staves, and is here worked over by the bare but cleanly-washed feet of the dairyman, or hired man. This working with the feet is continued, just as in kneading dough, till all is brought to a stiff paste.

When it has come to this consistence the forming of the cheese begins. The dairyman has for this purpose a cheese-mould standing before him, and lays on the bottom a layer of cheese without spice, and this is called the blind layer. The cheese tub or mould, Figs. 123 and 124, is used only for this first moulding. It is a wooden vat, made of staves from one to one and a half inches thick,

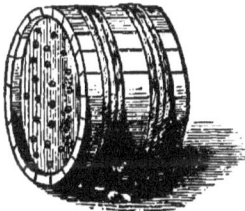

Fig. 123.

and is nine and a half to twelve and a half inches in diameter, and about ten inches high, bound at the bottom and top with stout hoops. The bottom of oak-wood, put in very carefully, is pierced with holes for letting off any moisture that may remain in the cheese. On the top of the tub a cover is exactly fitted, to sink down upon the cheese when the pressure is applied. This cover

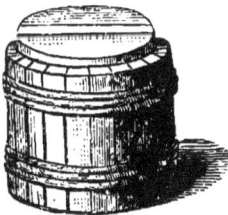

Fig. 124.

is of oak, one and a half inches thick, and has a cross-piece three and a half inches thick, which serves as a handle.

The first layer of cheese is quite firmly pressed down
or trodden into the mould with the hands or feet, and
then follows a layer of curd mixed with spices. The
mixture is made best by putting as much of the pasty
curd from the vat into a tub as will form one layer in
the mould.   Over this the spice is strewn, caraway and
some pounded cloves, and the mass is then worked over,
when it is placed as a new layer into the mould.   Upon
the second layer some coarsely-pounded cloves are
generally scattered, or they are stuck whole over the
surface.   After that the second layer is pressed in like
the first, and the third follows, and so on till the mould
is full.   On the uppermost and firmly pressed layer is
laid the cover.   The mould thus carefully filled is now
brought under a press, which, partly on account of its
length, is called the "long-press," and sometimes the
"first" or "cheese press," because the cheese first
comes under it. This press is seen in Fig. 125. It

Fig. 125.

stands on four short legs, and consists of upright beams
fixed upon a platform, and a long beam, acting as a lever,
with one end fastened by a rivet or bolt. The other
end is loaded with weights to any desirable extent, as
appears in the cut. The power of the press may also
be increased or diminished by shifting the end of the
lever to the lower or upper hole.

When the mould is put under the press it is set into a shallow, four-cornered wooden box or pan on the foot-board. This pan is furnished with grooves at the side, through which the whey can escape. The pressure may still further be increased by putting a block on the lid of the mould, as appears in the press. It is this power-ful pressure which gives the cheese the high quality for which it is distinguished above others. The whey still remaining in the curd runs off through the holes in the bottom of the mould, when the strong pressure is applied, into the pan, and is caught in another pan which sets under the press.

When the cheese has stood two hours under the press, it is taken from the mould, surrounded by a clean linen cloth, and again brought under the press. The change of cloth is repeated once or twice after two or

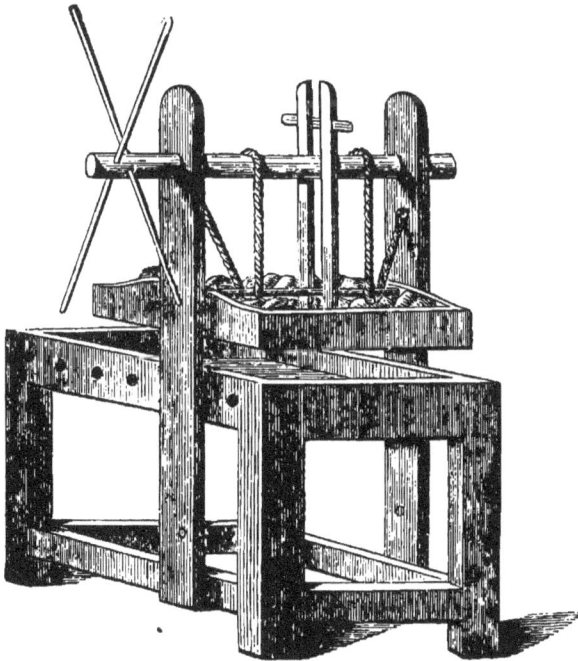

Fig. 120.

three hours' pressing, and the cheese is left standing in the press over night. The next morning the cheese is brought under another press, under which it is subjected to still more powerful pressure, and receives its peculiar form. This press is seen in Fig. 126, and consists of a frame resting on four strong uprights, forming a kind of firm table. On the plate of the table lie four or six rollers, whose ends at both sides pass through holes in the standard pieces, and serve merely to assist in taking out the cheese. The pressure is obtained by heavy weights let down and raised by a kind of wind-lass fixed in two perpendic-

Fig. 127.

ular standards. The cheese as it comes under this press is not in the mould, but is simply laid in a pan, as seen in Fig. 127. Before the pressure begins, however, the stamp or mark of the manufacturer, a key, a letter, etc., in iron, is laid upon the cheese, and upon that a square board. The pan and weight are lowered, so that the pressure begins and the stamp is impressed on the cheese, which becomes flatter, smoother, and firmer, than before. The cheese is left under this press till it gets its final form, and the pressure in the pan is increased or diminished, according to circumstances.

When the cheese, after being pressed in both ma-chines, has received its final form, it is placed in a long trough, called the salt-trough, which is generally in the cow-room behind the cow-stands. It has been already said that the cow-stall is used as a cheese-room in sum-mer, when the cows are out to pasture. In this trough, a space deep and wide enough for the diameter of the cheese, from four to six cheeses can be laid. In the salt-trough the cheeses are salted as long and as thor-oughly as is necessary. Observation and experience are needed here to get the right quantity of salt and the right time, that the cheese may receive a suitably firm crust or rind.

When the cheese in the salt-trough is sufficiently salted, it is put over a large tub, where it is properly washed in cold, fresh water, trimmed with a cheese-knife, and colored. For coloring, annatto boiled in water with some potash is used. After the coloring the cheese is rubbed with the beistings, or first milk of a cow newly-calved. The spice cheese gets its red color and firm, smooth rind in the coloring and washing in the beistings; and this distinguishes it from other sorts.

The colored cheeses are now laid upon shelves made for the purpose in the cow-stall used as a cheese-room, and turned daily till properly dried. When dry they are laid for sale in a cheese or store room. This room is connected with the house, or separated from the other rooms only by a thin board partition. This room, as well as the cow-stall, is kept extraordinarily clean,—scoured and aired, and used for nothing but the keeping of cheese.

Fig. 128 represents the cow-stall used as a cheese-room, in which the salt-trough is seen, and the dairyman and dairy-woman are occupied in turning and trimming the cheese.

MANUFACTURE OF SWEET MILK CHEESE IN SOUTH HOLLAND. — The best kind of sweet milk cheese is made in the vicinity of the city of Gouda, and on the gray and Dutch Yssel, from which circumstance it is often known by the name of Gouda cheese.

The making of this cheese is less difficult than that of spice cheese, but requires more attention and care, because the rich sweet milk is used for it. It is as follows: The milk as it comes fresh from the cow is strained through a hair-strainer into a large wooden vat

or tub, or, in some large dairies, into a copper kettle
which stands on a peculiar tray or bench.   This tray is
made of four to five inch posts, and its size is gov-

Fig. 128.

erned by the quantity of milk of the tubs to be used;
but these tubs generally hold from one hundred to one
hundred and fifty cans.   The milk is immediately set
with the requisite quantity of rennet, usually one quarter of a can to one hundred cans of milk; and if it does
not "come" in a quarter of an hour, more rennet is
added.

When it has properly curdled, it is stirred in all directions with a wooden ladle three or four times over, and

somewhat broken up, when it is allowed to stand three
or four minutes at rest. It is then gently and constantly
stirred again, with the ladle or the hands, and broken.
By too active stirring one gets more whey than cheese,
and very quick stirring must be avoided. The whey is
then allowed to stand some time, by which the curdled
cheese particles collect, and the whey appears on the
surface, and can be taken off and poured into a tub
made for the purpose. To the mass still remaining in
the kettle, which is now almost all cheesy matter, as
much hot water is added as is sufficient to warm it prop
erly. The addition of hot water must be made with
discretion, however, and must not exceed a certain
amount, which can be learned only by practice. The
more we add, the drier will the cheese become after a
while; and, though it may keep the better, and be better
for transportation, the taste is unquestionably injured by
it. The cold-made cheese is far more liable to injury
from keeping, but is much richer and more palatable,
on which account the best is generally eaten fresh. The
quantity of hot water to be added for warming the milk
must therefore be determined somewhat by the disposition to be made of the cheese.

When the hot water has stood, say half an hour, on
the curd, it is taken off and poured into the whey. The
curd is now properly brought together by the hands or
a ladle, and again thoroughly worked and broken. After
standing at rest a short time, the water and whey are
turned off again, as completely as possible, in the whey-
tub. The mass of curd still remaining in the vat, now
called wrongel, is cut up into small pieces, which are
very carefully worked over, and then pressed into the
wooden cheese-mould. In order to get a very fine separation of the curd, only a small quantity is taken at once
from the vat, which is rubbed in the hands, and then

29*

pressed into the mould till it is quite full. The cheese-mould is in the form of a bowl, made of willow wood, with its lower part pierced with holes, so that the whey can run off when the pressure is applied. The cheese now formed is taken out carefully, rubbed with the hands, and still further worked in the cheese-tub, and again very firmly pressed into the mould with the hands.

To be able to press it into the mould with greater power, an implement called the presser is used. It consists of a short stick, with a kind of handle or cross-piece on the upper end. On the lower end a disc is fixed which fits into the cheese-mould. In using the instrument, the disc is placed on the cheese to be pressed into the mould, the handle or cross-piece is placed against the chest or shoulders, and the operator presses down at the same time with his hands, thrusting the disc as deeply as possible into the cheese-mould. When pressed enough on one side, it is turned round in the mould, bringing the other side up, and the pressure is again applied as strongly as possible. For saving the whey in cheese-pressing, the mould is set into a pan only a little larger than the mould itself, which catches the whey running out from the mould. When the cheese in the mould is properly pressed by hand, the cover is put upon the mould, which is loaded gradually, in order to bring down the greatest possible pressure. The weight or pressure is greater or less according to the size of the cheese; yet during the pressure the cheese must be frequently turned, that it may get the right form. The gradual increase of the pressure goes on for twenty-four hours, when the cheese is taken from the mould to be laid in a tub of salt-brine in the cellar; the cellar must be kept cool. The cheese remains in the brine twenty-four hours, but is turned once in that time. It is then taken out and put upon a table, the surface of which is inclined, the legs of one end being longer than those of the other. On both sides of the inclined table run grooves in the direction of the inclination of the surface, which unite at the lower end, and serve as a way of escape for the brine or pickle into a tub below. Here the cheese is rubbed with salt, and a handful of salt is scattered over the top, when it is left standing for some time "in the salt." If one side was rubbed in the morning, it is turned at evening; and the other side is served in the same manner as the first. A cheese of from fifteen to sixteen pounds remains standing thus four or five days, according to the temperature. If the heat is great, it must stand the longer in the salt. When sufficiently salted, it is washed off in hot water, and taken to the cheese-room, where it is daily turned on dry, clean shelves. If it is still greasy or dauby on the outside, it is still further washed in water, and dried off with a coarse linen towel.

The cheese-room is generally kept closed by day to keep out the light and sun, which are not good for cheese. It is opened in the morning and evening to let in a little cooling air; yet a strong breeze is avoided by opening all the doors and windows at the same time, for the cheese will crack and break open if exposed to it.

Sweet milk cheese is fit for use at the age of four weeks. Strongly salted cheese does not ripen up so quickly as that which is salted less; but, if it takes longer, the loss is less, and, on that account, it is preferred for sending off to less salted cheese, which, on the other hand, is richer, and has a little better taste. In the daily turning of the cheese, great care is taken to observe any little specks in it where the mites conceal themselves. As soon as such places are discovered, a hole is dug out with a knife as deep as they extend into the cheese. The holes are left open till the next day, when, if no more mites appear, they are stopped up with other cheese. But, if they still appear, some pounded pepper is put into the holes, which destroys them. Rotten or moist spots on the cheese are treated in the same way, but very deep holes have to be made into the cheese, and it is best to cover them with buckwheat-meal, when they dry up very quickly. ɩ

In very hot weather it sometimes happens that the cheese swells up and begins to ferment. Then it is laid on the cleanly-scoured pavement of the cheese-room, where it is cooler; or, as many do, pierced pretty deeply with holes with a knitting-needle, which often helps it. With the decrease of the great heat of the sun, the swelling also ceases. The cheese is not injured except in appearance, the taste being improved. But, if the swelling is very considerable, it makes the cheese hollow. If the milk and cheese dishes are not very cleanly washed and rinsed out, the cheese gets a wrinkled crust, and begins to ferment.

Sweet milk cheese, three or four months old, is turned and aired only once a week in dry weather. Many cheese-makers also sprinkle the cheeses daily, for a week or two after they are fourteen days old, with beer and vinegar, or with vinegar in which saffron has been extracted, by which it gets not only a beautiful yellow color, but is also protected from flies.

THE USE OF THE WHEY OF SWEET MILK CHEESE. — On what remains of the milk devoted to the making of sweet milk cheese in the manner above described, or the whey which runs off in the pressing of the cheese, there forms, after it has stood a few days, a fine creamy skin, which is carefully taken off with a wooden spoon, put in a clean jar, and stirred from time to time. This cream is collected to make butter, and it can be done once a week. This butter-whey is healthful and good, to be sure ; but, on the whole, is not so fine and delicate flavored as good cream butter, and on this account is cheaper.

The butter-milk which comes from the churning of the cream of whey is a good food for swine. They greatly relish it.

Whey is also sold as a beverage, and is called "sweet whey." When fresh and untainted, it is quite an agreeable drink, very cooling, and good for the health in spring, purifying the blood, though somewhat purgative in its effect on the kidneys. Later in summer, when the heat is very great, whey is thought to be rather injurious to the health than otherwise. It is then used exclusively for swine.

MAY CHEESE. — In the early part of summer, when the grass is best, sweet milk cheese is made in precisely the same way as that described, yet of smaller size and less weight. This is called May cheese, and is designed for immediate use or sale when ripe, as it will not keep, and easily loses its fine flavor.

JEWS' CHEESE.—Another kind of sweet milk cheese is the Jews' cheese. It differs from common sweet milk cheese in its form, which is flatter and thinner, and partly in being less salted, and of a much looser texture. It is but little made ; but some dairies are devoted to it.

COUNCIL'S CHEESE. — This is made as the common sweet milk cheese, only in much smaller moulds. It has also a peculiar color. It is allowed to get rather old before it is relished, and is then mostly given away.

NEW MILK'S CHEESE. — This is made in winter, when the cows are in the stall. It is not so good as grass cheese, which is made in summer, when the cows are at pasture, and is less relished, and brings a lower price. When the cows are brought to the barn late in the fall, it can be made of very good quality for a few days ;

but the longer the cow remains in the stall the more
the milk loses its good quality for cheese, on which
account but few of the larger dairies make cheese at all
in winter.

To make it appear to buyers more like grass-made
cheese, and to be able to sell it, it is colored with the
same material, and it is then often very difficult to distinguish it, since great pains is taken to give the two
kinds the same form, hardness of rind, etc. The dairymen have less to do with this deception than the. dealers. Hay cheese is rather better in quality for coloring,
since it gains in appearance and taste ; but it never can
equal grass-made cheese in fine qualities.

CHEESE-MAKING IN NORTH HOLLAND. — In the province
of North Holland sweet milk cheese is made almost
exclusively. From ancient times this particular branch
of farming has been carried to great extent ; but it has
especially grown in importance since the province
gained a firm soil by artificial draining. At the present
time North Holland is the head-quarters of the cheese-
trade ; and it is easily explained in the fact that no
other province has more or better cattle. The manufacture of cheese is almost the only object of keeping
cattle, and the North Dutch dairy farmer applies himself with the greatest possible zeal to the most careful
modes of cheese-making, in order to keep up the ancient
reputation of his cheeses, both in the domestic and
foreign markets, and to secure to himself all of the
advantages springing from it.

The quantity of cheese which is weekly sold in the
markets of Alkmaar, Hoorn, Edam, Purmerend, Meden-
blik, Enkhuizen, etc., is enormous. We cite Alkmaar
alone as an example, where on the city scales there were
weighed no less than 23,859,258 Netherlandish pounds
(536,834,830 pounds, American), from 1758 to 1830.

Since that time the manufacture has increased, so that from three to four million Netherland pounds are annually brought to the Alkmaar market. But, besides this, a large quantity of cheese does not come into the market, but is sold at the dairy without passing through the hands of the traders, and never comes to the city scales.

In 1843 there were sold in the North Dutch cheese-markets 22,385,812 pounds, to say nothing of the large quantity sold directly from the dairy. It is easy to see, therefore, how important and extensive an interest the manufacture of cheese has become for this province. Of the twenty-two million pounds annually exported, the value may be estimated as at least three million Dutch guilders. The price and value of the cheese vary, of course, with the markets.

The North Dutch cheese differs somewhat in quality and money value, according to the section where it is made; but in general that made in the region about Hoorn is considered the best, as is very natural, since in that vicinity are to be found the finest meadows and pastures in the province. The villages of Ooster-blokker, Westerwoude, Hoogecarspel, and Twisk, are distinguished above all others; and so are the pastures of Beemster, Purmer, and Schermer, almost equally so.

The Dutch cheese-maker reckons twelve Nether-land cans of milk to a pound — two and a quarter pounds American — of cheese, according to which a cow in three hundred days would give from eighteen hundred to two thousand cans of milk, or usually from one hundred and fifty to one hundred and seventy-five Netherland pounds of cheese, in a year.

THE UTENSILS USED IN CHEESE-MAKING IN NORTH HOL-LAND are nearly the same as those already described for saving the milk for butter, and those used in the various processes of cheese-making in South Holland. They are modified to some extent, to be sure, by the taste, the pride, the wealth, or the caprice, of each dairy-man. Many of them are painted, wholly or in part, in oil colors, for the sake of durability as well as cleanli-ness, on which the North Dutch dairyman lays great stress. They do not require much capital.

VARIETY OF NORTH DUTCH CHEESES, AND THE TRADE IN THEM. — The North Dutch cheese is called sweet milk cheese, and also, pretty commonly, white cheese, where it is made ; but in Germany it is called Edamer. less because the best is made in the vicinity of this city than because the largest trade in it is carried on there.

All sweet milk cheese has not the same weight, form, and size. Many kinds of it come into the market under different names ; as, for example, large cheese of 20 to 24 pounds (45 to 54 pounds), Malbollen of 16 pounds (36 pounds), medium of 10 to 12 pounds (22 to 27 pounds), Commission's of 6 or 7 pounds (14 to 16 pounds), and little ones of 4 pounds (9 pounds), to which belong the Jews' cheese. Besides this, the making of English cheese is carried on. Malbollen is but little made. It is of about twenty pounds weight. Fifty years ago large quantities of it came into market, and were sold mostly in North Brabant and the Rhine provinces. Of the medium cheese the manufacture is pretty extensive at the present time, and it is sold to go to North Brabant chiefly. The price of these sorts is more frequently fluctuating than that of the smaller ones ; but less so than that of Commission's cheese, which is not much made. These varieties in former years were very profit-able, since they were made with little labor, being light and spongy from slight pressing and little salting, and were sold green.

Dairy industry is now chiefly devoted to making the varieties most known and sought for in Germany, the Edam small sweet milk cheeses, which are sent in enormous quantities to all parts of the world. There are two varieties of Edam cheese in the market, one with a white, the other with a red rind. The latter is firm, more of a yellowish color inside, and colored out. side. The coloring matter is prepared in France for this special purpose. By this treatment the cheese is better adapted to transportation. The early red rind cheese is the finest and best. It is made in spring from milk fresh and warm from cows just turned to pasture, and is exported mostly to Italy, Spain, and America. That made later in summer is not so good, and goes to France; the red rind, made still later in the fall, goes to England and Brabant. Cheese that is injured, or does not keep well, is sold mostly in Hamburg and Brabant.

MAKING OF EDAM CHEESE.—The Edam is a rich sweet milk cheese, that is made from fresh, unskimmed milk. The milk, while still warm from the cow, is poured into a large tub or a kettle through the strainer. In cold weather, when it has cooled off in standing in the air, it is warmed to a proper degree by adding milk heated by the fire. The rennet is then added. This is prepared in the following manner: The maw of the nursing-calf, cut into long strips, is soaked for twenty-four hours in sweet whey, when it is made lukewarm over a slow fire, whey and all, and three times the quantity of cheese-brine, or solution of the salt of the cheese, added. The mass is then allowed to stand four days, when it is fit for use. An exact determination of the quantity of rennet to be used cannot well be given, since the quantity depends on the quality; but usually about two hundred cans of milk to one fifth of a can.

30

of rennet is the proportion, taking more or less, accord
ing to the strength of the rennet.

The milk in the tub to which the rennet has been
added is covered over and allowed to stand till it is
curdled, or become hard, which usually requires a
quarter of an hour. The curdled milk is then called
" glib." It is now slowly but regularly stirred, with a
shallow, long-handled cheese-spoon, in all directions.

Some cheese-makers treat the milk in the following
manner : They stir the milk, thrusting an inverted
cheese-ladle into the curdling mass every two or three
minutes after adding the rennet, by which the curdling
is much hastened. Now they move the ladle or cheese-
stick three or four times with considerable force through
the thickening milk, and lay it, inverted, on the surface
of the milk, covering the vat for ten or twelve minutes,
when the mass is again set in motion, and then again
allowed to stand. By this means the cheese particles
settle to the bottom, and the whey rises to the top.

When, after these alternate stirrings and rest of the
curdling milk, the solid particles have settled, and the
whey is collected on top, the latter is turned off, as carefully as possible, into the whey-tub. In order the better
to settle the cheesy parts, and to cause the whey to come
up, the cheese-stick is loaded with weights or stones, by
which the whey is separated in the pressure upon the
curd. Some minutes after, the whey is again turned
off, the whole mass is properly stirred, and the curd is
collected with the cheese-stick and worked with the
hands, and the whey is again carefully turned off. The
curd, now become thick, is taken out of the vat, piece
by piece, and broken with the hands as finely as possible, in order to fill as much into the cheese-moulds as
will just make a cheese. The moulds are set into the
cheese-vat, and the curd is worked and pressed closely in

with the hand, to remove the whey as much as possible. The cheese is then taken out of the mould, and again very finely crumbled in the vat, and, after the whey is again turned off through the strainer, is pressed the second time into the mould, so that it is as full of cheese as it can possibly be. It is then turned in the mould so that the upper side goes down, when it is again firmly pressed in. The turning is repeated several times.

In the making of large and medium cheeses the presser is used, while space left empty by the pressure is again filled with curd, so that the mould is always full, and the cheese gets its requisite size. In the smaller or four-pound cheeses, the hands alone are used for this pressing into the mould. The mould, now pressed full, is put into a tub, properly washed in whey, and cleansed of all remaining fat. By the washing and smoothing the cheese must get a glossy and smooth rind. After this is done, the cheese is again taken out of the mould, wrapped in a clean linen cloth, put in again, and covered over and brought under the press, that it may become harder and firmer, and that the whey may run off.

In hot weather the cheese is left under the press five hours, from nine in the morning till two in the afternoon; but, if it is cool, it must stand longer. There are several different objects in view in deciding the continuance of the pressure. Many think two or three hours sufficient, whilst others press five hours. Cheese designed for export is pressed longer, or twelve hours.

It takes from three to four hours, usually, from the pouring in of the milk to the bringing of the cheese under the press; but it can be done in two or two and a half hours without injuring the cheese.

After the first pressing is finished, the cheese is put into another mould, rounder than the first, and with only one hole in the bottom, to lie in the salt. In many places a long trough is used, in which several such moulds are placed to be salted at the same time; and for this either dry salt or pickle (brine, or salt in solution) is used. The pickle is most commonly used, and is thought best. When one side of the cheese has laid some hours in the brine, it is turned, and the other side is also salted. After a while it is salted or turned in the brine but once a day. Small four-pound cheeses remain nine days in hot weather, and in cold ten or twelve days, in the salt; medium ones of ten to twelve pounds must lie at least three weeks. In very hot weather they are often salted twice a day. The moulds with the salted cheese are placed, several together, into the cheese-vat where the brine is, or on a salting-tray where the brine is collected in a tub beneath. After being finally salted, they are washed perfectly clean with water or warm whey. Many put their cheeses from the brine immediately in a kettle of hot whey for some minutes, and wash them in it. All unevenness or roughness got in pressing in the mould is now scraped off with a knife.

After the washing, the cheeses are again perfectly dried, and laid on the shelves in the cheese-room, where they are daily turned, and remain from two to four, and even five weeks. The cheese is now salable; but before it is packed or delivered it is laid for some hours to soak in pure, cold spring or well water, the smallest for three hours, the medium four, and the largest five hours. The cheese is then well cleaned with the cheese-brush, laid on the shelf in the store-room, and turned a week or more, daily. But, in order to give them a fine yellow color, in damp weather, especially, the poorer ones are, by many dairymen, laid a good ways apart, and sprinkled or washed daily with new beer. When the cheese is to be sold, it is properly washed still again in hot whey, and rubbed with a woolen cloth a day before sending to market, with hot or cold linseed-oil, by which the outside of the cheese gets a fine glow; but it must be rubbed till no fat or oil is to be felt.

THE RED COLOR OF EDAM CHEESE. — After the dairy-man has sold his cheese to the merchant, it is colored by him quite red. It will not be uninteresting to many readers to know some of the details of this peculiar color.

Edam cheese is colored with what is called tournesol, which is extracted from a plant (*Croton tinctorium*). This is an annual, which grows wild in France, in great abundance, in the vicinity of Montpelier, in Langue-doc; and around Aix, in Provence, large commons are sown with it. The seed is sown in March and April. From a white and straight tap root, it sends up a stalk something like six inches high, which divides into many branches. The leaves have very long stems, of a pale green color. The flower-stalks spring up from between the branches, and bear flowers in fan-shaped clusters. The vegetation of the plant continues four months.

The preparation of the tournesol is as follows: The plants are collected late in summer. the roots thrown away, and the other parts taken to a mill, where they are ground, and the juice pressed out. Into this juice the rags of old hempen cloth are dipped till they are soaked full, when they are hung up to dry in the sun. When they are dry they are laid on a tray over a tub filled with urine, in which carbonate of lime has been dissolved, so that the edges hang over the rim of the tub on which they rest. The vapor from the solu-tion of lime must penetrate the rags, and this gives them a violet color, when they are taken off and dried again, to be replaced till they are fully colored.

The tournesol rags have become an article of commerce, for which France receives annually from Holland from 100,000 to 200,000 guilders (from \$38,000 to \$76,000).

To give the Edam cheeses the red rind, they are rubbed with these tournesol rags, from which they get the dark violet color; and after they are dried they are again rubbed, which gives them a glowing red.

It is an excellent peculiarity of the tournesol rags that they not only impart the color to Edam cheese, to which people abroad are so accustomed, but that they keep the insects from the cheese, whilst the coloring matter does not penetrate inside, but remains on the rind. Substitutes for it have been repeatedly sought, but not found; nor have the attempts made to grow the plant in Holland proved successful.

USE OF THE WHEY OF THE NORTH DUTCH SWEET MILK CHEESE. — The whey obtained in making cheese in North Holland is collected in large tubs. The sweet, agreeable taste of the whey is soon lost when it is set to obtain the fatty particles still remaining in it. The cream which forms on it is daily taken off with a skimmer, put into a cream-pot, and when it is collected in sufficient quantity it is made into whey butter.

CHAPTER XII.

In the earlier chapters of this work I have spoken to farmers and dairymen of the selection, care, and management, of dairy stock. The seventh, eighth, and ninth chapters relate more especially to your depart- ment, and on your application and skill will depend chiefly the successful result of the dairy establishment. Of what avail are costly barns, well-selected cows, and judicious feeding, in the butter and cheese dairy, if the products are to be depreciated in value by the imper- fect modes of preparing them for the market, where the final judgment is passed upon them, and where it is expected the price will be according to their value?

You have, doubtless, had a much greater practical knowledge and experience of the details of dairy management than I have. For this practice and experi- ence I have the utmost respect; but I have not spoken without a knowledge of the subject. I have made many a cheese, and many a pound of butter, while my ob- servations have extended over all the most important dairy districts of the country, and have not been limited to the practices of any one section, which, however good in themselves, may not be the best. I trust, there- fore, you will excuse me for calling your attention to the more important points to which I have alluded; and, if my conclusions happen to differ from your own, in any respect, that you will not discard them as worthless, without first bringing them to the test of careful experiment, when I trust they will be found correct.

I have not written to establish any favorite theory, but simply to inculcate truth, and to aid in developing a most important branch of American industry, which, either directly or indirectly, involves the investment of a vast amount of capital, the aggregate profits of which depend so largely on your judgment and skill.

I need not remind you that any addition, however small, to the market value of each pound of butter or cheese, will largely increase the annual income of your establishment. Nor need I remind you that these articles are generally the last of either the luxuries or the necessaries of life in which city customers are willing to economize. They must and will have a good article, and are ready to pay for it in proportion to its goodness; or, if they desire to economize in butter, it will be in the quantity rather than the quality.

Poor butter is a drug in the market. Nobody wants it, and the dealer often finds it difficult to get it off his hands, when a delicate and finely-flavored article attracts attention and secures a ready sale. Some say that poor butter will do for cooking. But a good steak or mutton-chop is too expensive to allow any one to spoil it by the use of a poor quality of butter; and good pastry-cooks will tell you that cakes and pies cannot be made without good sweet butter, and plenty of it. These dishes relish too well, when properly cooked with nice butter, for any one to tolerate the use of poor butter in them.

On page 220 and elsewhere, I have dwelt on the necessity of extreme cleanliness in all the operations of the dairy; and this is the basis and fundamental principle of your business. I would not suppose, for a moment, that you are lacking in this respect. The enormous quantities of disgusting, streaky, and tallow-like butter that are daily thrust upon the seaboard markets must be due to the carelessness and negligence of heedless men, to exposure to sun and rain, to bad packing, and to delays in transportation. Many of these evils you may not be able to remove, since you cannot follow the article to the market, and see that it arrives safely and untainted. But you can take greater pains, perhaps, in some of the preliminary processes of making, and produce an article that will not be so liable to injure from keeping and transportation ; and then, if fault is to be found, it does not rest with you.

I will not suggest the possibility that your ideas of cleanliness and neatness may be at fault ; and that what may seem an excess of nicety and scrubbing to you may appear to be almost slovenliness to some others, whose butter receives the highest price in the market, and always finds the readiest sale. Permit me, however. to refer you to pages 300, 324, and 325, where a detailed account is given of the washings in water and washings in alkali ; of the scrubbings, and the scourings, and the scaldings, and the rinsings, which the neat and tidy Dutch dairy-women give all the utensils of the dairy, from the pails to the firkins and the casks, and also to their extreme carefulness that no infectious odor rises from the surroundings. I think you will see that it is a physical impossibility that any taint can affect the atmosphere or the utensils of such a dairy, and that many of the details of their practice may be worthy of imitation in our American dairies.

And here allow me to suggest that, though we may not approve of the general management in any particular section, or any particular dairy, it is rare that there is not something in the practice of that section that is really valuable and worthy of imitation.

On pages 231 and 234 I have called your attention to the use of the sponge and clean cloth for absorbing and removing the butter-milk in the most thorough manner ; this I regard as of great importance.

I have stated on page 234 that, under ordinarily favorable circumstances, from twelve to eighteen hours will be sufficient to raise the cream ; and that I do not believe it should stand over twenty-four hours under any circumstances. This, I am aware, is very different from the general practice over the country. But, if you will make the experiment in the most careful manner, setting the pans in a good, airy place, and not upon the cellar bottom, I think you will soon agree with me that all you get, after twelve or eighteen hours, under the best circumstances, or at most after twenty-four hours, will detract from the quality and injure the fine and delicate aroma and agreeable taste of the butter to a greater extent than you are aware of. The cream which rises from milk set on the cellar bottom acquires an acrid taste, and can neither produce butter of so fine a quality or so agreeable to the palate as that which rises from milk set on shelves from six to eight feet high, around which there is a full and free circulation of pure air. The latter is sweeter, and appears in much larger quantities in the same time than the former.

If, therefore, you devote your attention to the making of butter to sell fresh in the market, and desire to obtain a reputation which shall aid and secure the quickest sale and the highest price, you will use cream that rises first, and that does not stand too long on the milk. You will churn it properly and patiently, and not with too great haste. You will work it so thoroughly and completely with the butter-worker, and the sponge and cloth, as to remove every particle of butter-milk, never allowing your own or any other hands to touch it. You

will keep it at a proper temperature when making, and
after it is made, by the judicious use of ice, and avoid
exposing it to the bad odors of a musty cellar. You
will discard the use of artificial coloring or flavoring matter, and take the utmost care in every process of making. You will stamp your butter tastefully with some
mould which can be recognized in the market as yours;
as, for instance, your initials, or some form or figure
which will most please the eye and the taste of the
customer. You will send it in boxes so perfectly prepared and cleansed as to impart no taste of wood to the
butter. If all these things receive due attention, my
word for it, the initials or form which you adopt will
be inquired after, and you will always find a ready and
a willing purchaser at the highest market price.

But, if you are differently situated, and it becomes
necessary to pack and sell as firkin-butter, let me suggest the necessity of an equal degree of nicety and
care in preparation, and that you insist, as one of your
rights, that the article be packed in the best of oak-
wood firkins, thoroughly prepared after the manner of
the Dutch, as stated on page 325. A greater attention
to these points would make the butter thus packed
worth several cents a pound more when it arrives in
the market than it ordinarily is. Indeed, the manner
in which it not unfrequently comes to market is a disgrace to those who packed it; and it cannot be that
such specimens were ever put up by the hands of a
dairy-woman. I have often seen what was bought for
butter open so marbled, streaked, and rancid, that it was
scarcely fit to use on the wheels of a carriage.

If you adopt the course which I have recommended
in regard to skimming, you will have a large quantity
of sweet skimmed-milk, far better than it would be if
allowed to stand thirty-six or forty-eight hours, as is the

custom with many. This is too valuable to waste, and it is my opinion that you can use it to far greater profit than to allow it to be fed to swine. There can be no question, I think, that cheese-making should be carried on at the same time with the making of butter, in small and medium-sized dairies. You have seen, in Chapter XI., that some of the best cheese of Holland is made of sweet skim-milk. The reputation of Parmesan — a skim-milk cheese of Italy, page 266 — is world-wide, and it commands a high price and ready sale. The mode of making these varieties has been described in detail in the ninth and eleventh chapters ; and you can imitate them, or, perhaps, improve upon them, and thus turn the skim-milk to a very profitable account, if it is sweet and good. You will find, if you adopt this system, that your butter will be improved, and that, without any great amount of extra labor, you will make a large quantity of very good cheese, and thus add largely to the profit of your establishment, and to the comfort and prosperity of your family.

But, if you devote all your attention to the making of cheese, whether it is to be sold green, or as soon as ripe, or packed for exportation, I need not say that the same neatness is required as in the making of butter. You will find many suggestions in the preceding pages on the mode of preparation and packing, which I trust will prove to be valuable and applicable to your circumstances. There is a general complaint among the dealers in cheese that it is difficult to get a superior article. This state of things ought not to exist. I hope the time is not far distant when a more general attention will be paid to the details of manufacture, and let me remind you that those who take the first steps in improvement will reap the greatest advantages.

CHAPTER XIII.

THE PIGGERY AS A PART OF THE DAIRY ESTAB-
LISHMENT.

THE keeping of swine is incidental to the well-managed dairy, and both the farmer and the dairyman unite
it, to some extent, with other branches of farming.

In the regular operations of the dairy, however economically conducted, there will always be more or less
refuse in the shape of whey, butter-milk, or skim-milk.
which may be consumed with profit by swine, and
which might otherwise be lost. Dairy-fed pork is distinguished for its fineness and delicacy; and the dairy
refuse, in connection with grains, potatoes, and scraps,
is highly nutritious and fattening.

There is a wide difference between the profit to be
derived from the different breeds. Some are far more
thrifty than others, and arrive at maturity earlier. But
the choice of a breed will depend, to considerable
extent, on the locality and the object in view, whether
it be to breed for sale as stock, or for pork or bacon.

To get desirable crosses, some breeds must be kept
pure, especially in the hands of stock breeders, or those
who raise to sell as pure-bred, even though as pure
breeds they may not be most profitable to the practical
farmer and dairyman. Those who confine themselves
to the pure breeds, therefore, do good service to the
community of farmers and dairymen, who can avail
themselves of the results of their experience and skill.

31

I think it will generally be conceded that the size of the male is of less importance than his form, his tendency to lay on large amounts of fat in proportion to the food he eats, or his early maturity. Smallness of bone and compactness of form indicate early maturity; and this is an essential element in the calculations of the dairy farmer, who generally raises for pork rather than for bacon, and whose profit will consist in fattening and turning early, or, at most, as young as from twelve to fifteen months. A fine and delicate quality of pork is at the present time highly prized in the markets, and commands the highest price. For bacon, a much larger hog is preferred; but there can be little doubt that the cross of the pure Suffolk or Berkshire boar and the large, heavy and coarse sow, not uncommon in the Western States, would produce an offspring far superior to the class of hogs usually denominated "subsoilers," with their long and pointed snouts, and their thin, flabby sides. The principles of breeding, as stated on pp. 70 and 71, and elsewhere in the preceding pages, are equally applicable here, and are abundantly suggestive on many other points. This is the important point, the selection of the proper breed and the proper cross: for there is scarcely any class of stock which varies so much in its net returns as this; and there is none which, if properly selected and judiciously managed, returns the investment so quickly.

Those who feed for the early market, and desire to realize the largest profits with the least outlay of time and money, will resort to the Suffolk, the Berkshire, or the Essex, to obtain crosses with sows of the larger breeds, and will breed up more or less closely to these breeds, according to the special object they have in view. The Suffolks are nearly allied to the Chinese, and possess much the same characteristics. Though generally regarded as too small for profit except to those who breed for stock, their extraordinary fattening qualities and their early maturity adapt them eminently for crossing with the larger breeds. The form of the well-built Suffolk, when not too closely inbred, is a model of compactness, and lightness of bone and offal. Though often too short in the body, a large-boned female will generally correct this fault, and produce an offspring suited to the wants of the dairy farmer.

The Berkshire is also mixed in with the Chinese, and owes no small part of its valuable characteristics to that race. The Berkshires, as a breed, often attain considerable size and weight.

The improved Essex are the favorites of some, and for early maturity they are difficult to surpass. Some think they require greater care and better feeding than the Berkshire.

What is wanted is to unite, so far as possible, the early maturity and the facility to take on fat of the Suffolk, the Chinese, or the Essex, with a tendency at the same time to make flesh as well as fat; or, in other words, to attain a good growth and size, and to fatten easily when the time comes to put them down. The Chinese or the Suffolk are but ill adapted for hams and bacon; but, crossed upon the kind of hog already described, the produce will be likely to be valuable.

The most judicious practical farmers are now fully satisfied, I think, that the tendency, for the last ten years, in the Eastern States more especially, has been to breed too fine; and that the result of this error has been to cover our swine with fat at a very early age, and before they have attained a respectable size. In other words, the flesh and bone have been too far sacrificed to fat. A reaction has already taken place in the opinions on this point, and perhaps some caution may be necessary, that it does not lead too far in the opposite direction.

Some practical dairymen think that with a dairy of twenty or thirty cows they can keep from forty to fifty swine, by turning into the orchard or the pasture, in early spring, and as pigs, where they will easily procure a large part of their food, till the close of fall, when they are taken in and fed up gradually at first, but afterward more highly, and fattened as rapidly and turned as soon as possible.

Others say there is no profit in working hogs, and that they should be kept confined and constantly and rapidly growing up to the time of turning them for pork, growing steadily, but not laying on too much fat till fed up to it.

I am inclined to think the farmers of the Eastern States confine their swine too closely; and that, while still kept as store-pigs, a somewhat greater range in the orchard, or the pasture, would prove to be good economy, particularly up to the age of eight or nine months.

The judicious dairyman will study the taste and demands of the market where his pork is to be sold. If he supplies a city customer, he knows he must raise a fine and delicate quality of pork; and to do this he must select stock that will early arrive at maturity, and that will bear forcing ahead and selling young. If he supplies a market where large amounts of pork are salted and packed for shipping, or for bacon, a larger and coarser hog, fed to greater age and weight, will turn to better advantage, though I think a strain of finer blood will even then be profitable to the feeder. In either case, the refuse of the dairy is of considerable value, and should be saved with scrupulous care, and judiciously fed. "Many a little makes a mickle."

APPENDIX.

THE following is Mr. Thomas Horsfall's statement, referred to on page 138, with the omission of a few passages, relating to matters not immediately connected with the dairy. It is entitled

THE MANAGEMENT OF DAIRY CATTLE.

ON entering upon a description of my treatment of cows for dairy purposes, it seems pertinent that I should give some explanation of the motives and considerations which influence my conduct in this branch of my farm operations.

I have found it stated, on authority deserving attention, that store cattle of a fair size, and without other occupation, maintain their weight and condition for a length of time, when supplied daily with one hundred and twenty pounds of Swedish turnips and a small portion of straw. The experience of the district of Craven, in Yorkshire, where meadow hay is the staple food during winter, shows that such cattle maintain their condition on one and a half stone, or twenty-one pounds, of meadow hay each per day. These respective quantities of turnips and of hay correspond very closely in their nutritive properties; they contain a very similar amount of albuminous matter, starch, sugar, etc., and also of phosphoric acid. Of oil — an important element, especially for the purpose of which I am treating — the stated supply of meadow hay contains more than that of turnips. If we supply cows in milk, of

31*

average size, with the kind and quantity of food above mentioned, they will lose perceptibly in condition. This is easily explained when we find their milk rich in substances which serve for their support when in store condition, and which are shown to be diverted in the secretion of milk.

In the neighborhood of towns where the dairy produce is disposed of in new milk, and where the aim of dairymen is to produce the greatest quantity, too frequently with but little regard to quality, it is their common practice to purchase incalving cows. They pay great attention to the condition of the cow; they will tell you, by the high comparative price they pay for animals well stored with flesh and fat, that condition is as valuable for them as it is for the butcher; they look upon these stores as materials which serve their purpose; they supply food more adapted to induce quantity than quality, and pay but little regard to the maintenance of the condition of the animal. With such treatment, the cow loses in condition during the process of milking, and when no longer profitable is sold to purchasers in farming districts where food is cheaper, to be fattened or otherwise replenished for the use of the dairy keeper. We thus find a disposition in the cow to apply the aliment of her food to her milk, rather than to lay on flesh or fat; for not only are the elements of her food diverted to this purpose, but, to all appearance, her accumulated stores of flesh and fat are drawn upon, and converted into components of milk, cheese, or butter.

As I am differently circumstanced, — a considerable portion of my dairy produce being intended for butter, for which poor milk is not adapted,—and as I fatten not only my own cows, but purchase others to fatten in addition, I have endeavored to devise food for my milch cows adapted to their maintenance and improvement, and with this view I have paid attention to the composition of milk. From several analyses I have selected one by Haidlen, which I find in publications of repute. Taking a full yield of milk, four gallons per day, which will weigh upwards of forty pounds, this analysis assigns to it of dry material 5.20, of which the proportion, with sufficient accuracy for my purpose, consists of

Pure caseine, ...	2.00	pounds.
Butter, ...	1.25	"
Sugar, ...	1.75	"
Phosphate of lime, ...	.09	"
Chloride of potassium, ...		
Other mineral ingredients, ...	.11	"
	5.20	"

It appeared an object of importance, and one which called for my particular attention, to afford an ample supply of the elements of food suited to the maintenance and likewise to the produce of the animal; and that, if I omitted to effect this, the result would be imperfect and unsatisfactory. By the use of ordinary farm produce only, I could not hope to accomplish my purpose. Turnips are objectionable on account of their flavor; and I seek to avoid them as food for dairy purposes. I use cabbages, kohl rabi, and mangold wurzel, yet only in moderate quantities. Of meadow hay it would require, beyond the amount necessary for the maintenance of the cow, an addition of fully twenty pounds for the supply of caseine in a full yield of milk (sixteen quarts); forty pounds for the supply of oil for the butter, whilst nine pounds seem adequate for that of the phosphoric acid. You cannot, then, induce a cow to consume the quantity of hay requisite for her maintenance, and for a full yield of milk of the quality instanced. Though it is a subject of controversy whether butter is wholly derived from vegetable oil, yet the peculiar adaptation of this oil to the purpose will, I think, be admitted. I had, therefore, to seek assistance from what are usually termed artificial feeding substances, and to select such as are rich in albumen, oil, and phosphoric acid; and I was bound also to pay regard to their comparative cost, with a view to profit, which, when farming is followed as a business, is a necessary, and in any circumstances an agreeable accompaniment.

I think it will be found that substances peculiarly rich in nitrogenous or other elements have a higher value for special than for general purposes, and that the employment of materials characterized by peculiar properties for the attainment of special objects has not yet gained the attention to which it is entitled.

I have omitted all reference to the heat-supplying elements — starch, sugar, etc. As the materials commonly used as food for cattle contain sufficient of these to effect this object, under exposure to some degree of cold, I have a right to calculate on a less consumption of them as fuel, and consequently a greater surplus for deposit as sugar, and probably also as fat, in consequence of my stalls being kept during winter at a temperature of nearly sixty degrees.

The means used to carry out his objects are stated on page 138.

As several of these materials — rape-cake, shorts, bean-straw, etc. — are not commonly used as food, I may be allowed some observations on their properties. Bean-straw uncooked is dry and unpalatable. By the process of steaming, it becomes soft and pulpy, emits an agreeable odor, and imparts flavor and relish to the mess. For my information and guidance I obtained an analysis of bean-straw of my own growth, on strong and high-conditioned land; it was cut on the short side of ripeness, but yielding a plump bean. The analysis by Professor Way shows a percentage of

Moisture,	. . . . .	14.47	Woody fibre, ...	25 84
Albuminous mater,	. .	16.38	Starch, gum, etc., . . .	31.63
Oil or fatty matter,	. .	2.23	Mineral matters, . . .	9.45

Total, ... 100.00

In albuminous matter, which is especially valuable for milch cows, it has nearly double the proportion contained in meadow hay. Bran also undergoes a great improvement in its flavor by steaming, and it is prob-ably improved in its convertibility as food. It contains about fourteen per cent. of albumen, and is peculiarly rich in phosphoric acid, nearly three per cent. of its whole substance being of this material. The properties of rape-cake are well known: the published analyses give it a large proportion (nearly thirty per cent.) of albumen; it is rich in phosphates, and also in oil. This is of the unctuous class of vegetable oils, and it is to this property that I call particular attention. Chemistry will assign to this material, which has hitherto been comparatively neglected for feeding, a first place for the purpose of which I am treating. If objection should occur on account of its flavor, I have no diffi-culty in stating that by the preparation I have described I have quite overcome this. I can easily persuade my cattle (of which sixty to eighty pass through my stalls in a year), without exception, to eat the requisite quantity. Nor is the flavor of the cake in the least perceptible in the milk or butter.

During May, my cows are turned out on a rich pas-ture near the homestead; towards evening they are again housed for the night, when they are supplied with a mess of the steamed mixture and a little hay each morning and evening. During June, when the grasses are better grown, mown grass is given to them instead of hay, and they are also allowed two feeds of steamed mixture. This treatment is continued till October, when they are again wholly housed.

The results which I now proceed to relate are de-rived from observations made with the view of enabling me to understand and regulate my own proceedings.

GAIN OR LOSS OF CONDITION ASCERTAINED BY WEIGH-ING CATTLE PERIODICALLY. — For some years back I have regularly weighed my feeding stock, a practice from which I am enabled to ascertain their doings with greater accuracy than I could previously. In January, 1854, I commenced weighing my milch cows. It has been shown, by what I have premised, that no accurate estimate can be formed of the effect of the food on the production of milk, without ascertaining its effect on the condition of the cows. I have continued the practice once a month, almost without omission, up to this date. The weighings take place early in the morning, and before the cows are supplied with food. The weights are registered, and the length of time (fifteen months) during which I have observed this practice enables me to speak with confidence of the results.

The cows in full milk, yielding twelve to sixteen quarts each per day, vary but little; some losing, others gaining, slightly; the balance in the month's weighing of this class being rather to gain. It is common for a cow to continue a yield from six to eight months before she gives below twelve quarts per day, at which time she has usually, if not invariably, gained weight.

The cows giving less than twelve quarts and down to five quarts per day are found, when free from ailment, to gain, without exception. This gain, with an average yield of nearly eight quarts per day, is at the rate of seven pounds to eight pounds per week each.

My cows in calf I weigh only in the incipient stages; but they gain perceptibly in condition, and consequently in value. They are milked till within four weeks to five weeks previous to calving. I give the weights of three of these, and also of one heifer, which calved in March, 1855:

No.			1854.				1855.			Gain
				cwt.	qr.	lbs.		cwt.	qr.	lbs.	lbs
1	Bought and weighed,	July.	10	1	20	April.	11	3	0	148
2	" " "	"	8	2	10	"	10	2	0	214
3	" " "	"	8	2	0	"	10	0	0	184
4	Heifer, which calved also in March, 1855, weighed	"	7	0	0	"	9	3	0	300

These observations extend over lengthened periods, on the same animals, of from thirty to upwards of fifty weeks. A cow, free from calf, and intended for fatten ing, continues to give milk from ten months to a year after calving, and is then in a forward state of fatness requiring but a few weeks to finish her for sale to the butchers.

It will thus appear that my endeavors to provide food adapted to the maintenance and improvement of my milch cows have been attended with success.

On examining the composition of the ordinary food which I have described, straw, roots, and hay, it appears to contain the nutritive properties which are found adequate to the maintenance of the animal, whereas the yield of milk has to be provided for by a supply of extra food ; the rape-cake, bran, and bean-meal, which I give, will supply the albumen for the caseine ; it is somewhat deficient in oil for the butter, whilst it will supply in excess the phosphate of lime for a full yield of milk. If I take the class of cows giving less than twelve quarts per day, and take also into account a gain of flesh of seven to nine pounds per week, though I reduce the quantity of extra food by giving less of the bean-meal, yet the supply will be more in proportion than with a full yield ; the surplus of nitrogen and phosphoric acid, or phosphate of lime, will go to enrich the manure.

I cannot here omit to remark on the satisfaction I derive from the effects of this treatment on the fertility of the land in my occupation. My rich pastures are not tending to impoverishment, but to increased fertility ; their improvement in condition is apparent. A cow in full milk, giving sixteen quarts per day, of the quality analyzed by Haidlen, requires, beyond the food necessary for her maintenance, six to eight pounds per day of substances containing thirty or twenty-five per cent. of protein. A cow giving on the average eight quarts per day, with which she gains seven to nine pounds per week, requires four to five pounds per day of substances rich in protein, beyond the food which is necessary for her maintenance. Experience of fattening gives two pounds per day, or fourteen pounds per week, as what can be attained on an average, and for a length of time. If we considered half a pound per day as fat, which is not more than probable, there will be one and a half pounds for flesh, which, reckoned as dry material, will be about one third of a pound, which is assimilated in increase of fibrin, and represents only one and one third to two pounds of substances rich in protein, beyond what is required for her maintenance.

If we examine the effects on the fertility of the land, my milch cows, when on rich pasture, and averaging a yield of nine quarts per day, and reckoning one cow to each acre, will carry off in twenty weeks twenty-five pounds of nitrogen, equal to thirty of ammonia. The same quantity of milk will carry off seven pounds of phosphate of lime in twenty weeks from each acre.

A fattening animal, gaining flesh at the rate I have described, will carry off about one third of the nitrogen (equal to about ten pounds of ammonia) abstracted by the milch cow, whilst if full grown it will restore the whole of the phosphate.

It is worthy of remark that experience shows that rich pastures, used for fattening, fully maintain their fertility through a long series of years, whilst those used for dairy cows require periodical dressings to preserve their fertility.

If these computations be at all accurate, they tend to show that too little attention has been given to the supply of substances rich in nitrogenous compounds in the food of our milch cows, whilst we have laid too much stress on this property in food for fattening cattle. They tend also to the inference that in the effects on the fertility of our pastures used for dairy purposes we derive advantage not only from the phosphate of lime, but also from the gelatine of bones used as manure.

On comparing the results from my milch cows fed in summer on rich pasture, and treated at the same time with the extra food I have described, with the results when on winter food, and whilst wholly housed, taking into account both the yield of milk and the gain of weight, I find those from stall-feeding full equal to those from depasture. The cows which I buy as strippers, for fattening, giving little milk, from neighboring farmers who use ordinary food, such as turnips with straw or hay, when they come under my treatment increase their yield of milk, until after a week or two they give two quarts per day more than when they came, and that too of a much richer quality.

RICHNESS OF MILK AND CREAM.—I sometimes observe, in the weekly publications which come under my notice, accounts of cows giving large quantities of butter. These are usually, however, extraordinary instances, and not accompanied with other statistical information requisite to their being taken as a guide ; and it seldom happens that any allusion is made to the effects of the food on the condition of the animals, without which no accurate estimate can be arrived at. On looking over several treatises to which I have access, I find the following statistics on dairy produce : Mr. Morton, in his " Cyclopædia of Agriculture," p. 621, gives the results of the practice of a Mr. Young, an extensive dairy-keeper in Scotland. The yield of milk per cow is stated at six hundred and eighty gallons per year ; he obtains from sixteen quarts of milk twenty ounces of butter, or for the year two hundred and twenty-seven pounds per cow; from one gallon of cream three pounds of butter, or twelve ounces per quart (wine measure). Mr. Young is described as a high feeder ; linseed is his chief auxiliary food for milch cows. Professor Johnston (" Elements of Agricultural Chemistry") gives the proportion of butter from milk at one and a half ounces per quart, or from sixteen quarts twenty-four ounces, being the produce of four cows of different breeds, — Alderney, Devon, and Ayrshire,— on pasture, and in the height of the summer season. On other four cows of the Ayrshire breed he gives the proportion of butter from sixteen quarts as sixteen ounces, being one ounce per quart. These cows were likewise on pasture. The same author states the yield of butter as one fourth of the weight of cream, or about ten ounces per quart. Mr. Rowlandson (" Journal of the Royal Agricultural Society," vol. xiii., p. 38) gives the produce of 20,110 quarts of milk churned by hand as 1109 pounds of butter, being at the rate of fully 14 ounces per 16 quarts of milk ; and from 23,156 quarts of milk 1525 pounds

32

of butter, being from 16 quarts nearly 16¾ ounces of butter. The same author states that the yield of butter derived from five churnings, of 15 quarts of cream each, is somewhat less than 8 ounces per quart of cream. Dr. Muspratt, in his work on the "Chemistry of Arts and Manufactures," which is in the course of publication, gives the yield of butter from a cow per year in Holstein and Lunenburg at 100 pounds, in England at 160 pounds to 180 pounds. The average of butter from a cow in England is stated to be eight or nine ounces per day, which, on a yield of eight to nine quarts, is one ounce per quart, or for sixteen quarts sixteen ounces. The quantity of butter derived from cream is stated as one fourth, which is equal to about nine ounces per quart. The richest cream of which I find any record is that brought to the Royal Society's meeting during the month of July, for the churns which compete for the prize. On referring to the proceedings of several meetings, I find that fourteen ounces per quart of cream is accounted a good yield.

I have frequently tested the yield of butter from a given quantity of my milk. My dairy produce is partly disposed of in new milk, partly in butter and old milk, so that it became a matter of business to ascertain by which mode it gave the best return. I may here remark that my dairy practice has been throughout on high feeding, though it has undergone several modifications. The mode of ascertaining the average yield of butter from milk has been to measure the milk on the churning-day, after the cream has been skimmed off, then to measure the cream, and having, by adding together the two measurements, ascertained the whole quantity of milk (including the cream), to compare it with that of the butter obtained. This I consider a more accurate method than measuring the new milk, as there is a considerable escape of gas, and consequent subsidence, whilst it is cooling. The results have varied from twenty-four to twenty-seven and a quarter ounces from sixteen quarts of milk. I therefore assume in my calculation sixteen quarts of milk as yielding a roll (twenty-five ounces) of butter.

As I have at times a considerable number of cows bought as strippers, and fattened as they are milked, which remain sometimes in my stalls eight or nine months, and yield towards the close but five quarts per day, I am not enabled to state with accuracy and from ascertained data the average yield per year of my cows kept for dairy purposes solely. However, from what occurs at grass-time, when the yield is not increased, and also from the effects of my treatment on cows which I buy, giving a small quantity, I am fully persuaded that my treatment induces a good yield of milk.

As the yield of butter from a given quantity of cream is not of such particular consequence, I have not given equal attention to ascertain their relative proportions. I have a recollection of having tested this on a former occasion, when I found fourteen to sixteen ounces per quart, but cannot call to mind under what treatment this took place.

On questioning my dairy-woman, in December, 1854, as to the proportion of cream and butter, she reported nearly one roll of twenty-five ounces of butter to one quart of cream. I looked upon this as a mistake. On its accuracy being persisted in, the next churning was carefully observed, with a like proportion. My dairy cows averaged then a low range of milk as to quantity— about eight quarts each per day. Six of them, in a forward state of fatness, were intended to be dried for finishing off in January ; but, owing to the scarcity and consequent dearness of calving cows, I kept them on in milk till I could purchase cows to replace them, and it was not till February that I had an opportunity of doing so. I then bought four cows within a few days of calving ; they were but in inferior condition, and yielded largely of milk. Towards the close of February and March, four of my own dairy cows, in full condition, likewise calved. During March, three of the six which had continued from December, and were milked nearly up to the day of sale, were selected by the butcher as fit for his purpose. Each churning throughout was carefully observed, with a similar result, varying but little from twenty-five ounces of butter per quart of cream ; on Monday, April 30, sixteen quarts of cream having yielded sixteen rolls (of twenty-five ounces each) of butter. Though I use artificial means of raising the temperature of my dairy, by the application of hot water during cold weather, yet, my service-pipes being frozen in February, I was unable to keep up the temperature, and it fell to forty-five degrees. Still my cream, though slightly affected, was peculiarly rich, yielding twenty-two ounces of butter per quart. Throughout April the produce of milk from my fifteen dairy cows averaged full one hundred and sixty quarts per day.

My cows are bought in the neighboring markets with a view to their usefulness and profitableness. The breeds of this district have a considerable admixture of the short-horn, which is not noted for the richness of its milk. It will be remarked that during the time these observations have been continued on the proportion of butter from cream, more than half of my cows have been changed.

Having satisfied myself that the peculiar richness of my cream was due mainly to the treatment of my cows which I have sought to describe, it occurred to me that I ought not to keep it to myself, inasmuch as these results of my dairy practice not only afforded matter of interest to the farmer, but were fit subjects for the investigation of the physiologist and the chemist. Though my pretensions to acquirements in their instructions are but slender, they are such as enable me to acknowledge benefit in seeking to regulate my proceedings by their rules.

In taking off the cream I use an ordinary shallow skimmer of tin perforated with holes, through which any milk gathered in skimming escapes. It requires care to clear the cream ; and even with this some streakiness is observable on the surface of the skimmed milk. The milk-bowls are of glazed brown earthen ware, common in this district. They stand on a base of six to eight inches, and expand at the surface to nearly twice that width. Four to five quarts are contained in each bowl, the depth being four to five inches at the centre. The churn I use is a small wooden one, worked by hand, on what I believe to be the American principle. I have forwarded to Professor Way a small sample of butter for analysis; fifteen quarts of cream were taken out of the cream-jar, and churned at three times in equal portions:

The first five quarts of cream gavé .	. 127 ounces of butter.
Second five	"	"	"	"	. 125 " " "
Third five	"	"	"	"	. 120½ " " "

$$\overline{372\tfrac{1}{2}}$$

Equal to 24¾ ounces per quart.

At a subsequent churning of fourteen quarts of cream,

The first seven gave 7 rolls, or . . . 175 ounces of butter.
Second seven gave 7 rolls 2 oz., or . . 177 " " "

$$\overline{352}$$

Equal to 25¼ ounces per quart.

On testing the comparative yield of butter and of butter-milk, I find seventy per cent. of butter to thirty per cent. of butter-milk, thus reversing the proportions given in the publications to which I have referred. An analysis of my butter by Professor Way gives:

Pure fat or oil, ...	82.70
Caseine or curd, ...	2.45
Water, with a little salt, ...	14.85
Total, ...	100.00

The only analyses of this material which I find in the publications in my hand are two by Professor Way, "Journal," vol. xi., p. 735, "On butter by the common and by the Devonshire method;" the result in one hundred parts being:

	Raw.	Scalded.
Pure butter, ...	79.72	79.12
Caseine, &c., ...	3.38	3.37
Water, ...	16.90	17.51
Total, ...	100.00	100.00

32*

The foregoing observation of dairy results was con-tinued up to grass time in 1855. In April and May the use of artificial means was discontinued, without dimi-nution in the yield of butter or richness of cream, the natural temperature being sufficient to maintain that of my dairy at 54° to 56°.

I now proceed to describe the appearances since that time. In the summer season, whilst my cows were grazing in the open pastures during the day and housed during the night, being supplied with a limited quantity of the steamed food each morning and evening, a marked change occurred in the quality of the milk and cream; the quantity of the latter somewhat increased, but, instead of twenty-five ounces of butter per quart of cream, my summer cream yielded only sixteen ounces per quart.

I would not be understood to attribute this variation in quality to the change of food only. It is commonly observed by dairy-keepers that milk, during the warm months of summer, is less rich in butter, owing probably to the greater restlessness of the cows, from being teased by flies, etc. I am by no means sure that, if turning out during the warm months be at all advisable, it would not be preferable that this should take place during the night instead of during the day time. Towards the close of September, when the temperature had become much cooler, and the cows were supplied with a much larger quantity of the steamed food, results appeared very similar to those which I had observed and described from December to May, 1855. During the month of November the quality was tested with the following result:

From two hundred and fifty-two quarts of old milk were taken twenty-one quarts of cream, of which twenty were churned, and produced four hundred and sixty-eight ounces of butter, which shows:

27.50 ounces of butter from 16 quarts of new milk.
23.40   "   "   "   "   each quart of cream.

During May, 1856, my cows being on open pasture

during the day were supplied with two full feeds of the steamed mixture, together with a supply of green rape-plant each morning and evening.

The result was that from three hundred and twenty-four quarts of old milk twenty-three quarts of cream were skimmed, of which twenty-two were churned, and produced five hundred and fifteen ounces of butter, which shows :

24   ounces of butter from 16 quarts of new milk.
22.41   "   "   "   "   each quart of cream.

There is, doubtless, some standard of food adapted to the constitution and purposes of animals, combining with bulk a due proportion of elements of respiration, such as sugar, starch, &c., together with those of nutrition, namely, nitrogenous compounds, phosphates, and other minerals; nor can we omit oil or fat-forming substances ; for, however we may be disposed to leave to philosophy the discussion as to whether sugar, starch, &c., are convertible into fat, yet I think I shall not offend the teacher of agricultural chemistry by stating that the more closely the elements of food resemble those in the animal and its product, the more efficacious will such food be for the particular purpose for which it is used.

Sugar, starch, &c., vary very considerably in form and proportion from vegetable oils, which closely resemble animal fats.

When we consider that plants have a two-fold function to perform,— namely, to serve as food for animals, and also for the reproduction of the like plants, — and that, after having undergone the process of digestion, they retain only one half or one third of their value as manure, the importance of affording a due but not excessive supply of each element of food essential to the wants and purposes of the animal will be evident. If we fall short, the result will be imperfect; if we supply in excess, it will entail waste and loss.

Linseed and rape-cake resemble each other very closely in chemical composition ; the latter is chiefly used for manure, and its price ranges usually about half that of linseed-cake. In substances poorer in nitrogen, and with more of starch, gum, oil, &c., the disparity in value as food and as manure will be proportionately greater.

During the present season, Mr. Mendelsohn, of Berlin, and Mr. Gausange, who is tenant of a large royal domain near Frankfort on the Oder, on which he keeps about one hundred and fifty dairy cows, have been my visitors. These gentlemen have collected statistics in dairy countries through which they have travelled. I learned from them that in Mecklenburg, Prussia, Holland, &c., fourteen quarts of milk yield, on the average, one pound of butter; in rare instances twelve quarts are found to yield one pound. Both attach great importance to the regulation of the temperature. Mr. Mendelsohn tells me that the milk from cows fed on draff (distillers' refuse) requires a higher temperature to induce its yield of butter than that from cows supplied with other food.

On inquiry in my own neighborhood, I find it is computed that each quart at a milking represents one pound of butter per week. Thus, a cow which gives four quarts at each milking will yield in butter four pounds per week, or from fifty-six quarts sixty-four ounces of butter, or from fourteen quarts of milk one pound of butter. Taking the winter produce alone, it is lower than this; the cream from my neighbors' cows, who use common food, hay, straw, and oats, somewhat resembles milk in consistence, and requires three to four hours, sometimes more, in churning. On one occasion, a neighboring dairy-woman sent to borrow my churn, being unable to make butter with her own; I did not inquire the result. If she had sent her cow, I could in the course of a week have insured her cream which would make butter in half an hour. These dairy people usually churn during winter in their kitchen, or other room with a fire. Each of them states that from bean or oat meal used during winter as an auxiliary food they derive a greater quantity of butter, whilst those who have tried linseed-oil have perceived no benefit from it.

My own cream during the winter season is of the

consistence of paste, or thick treacle. When the jar is full, a rod of two feet long will, when dipped into the cream to half its length, stand erect. If I take out a teacupful in the evening, and let it stand till next morning, a penny-piece laid on its surface will not sink; on taking it off, I find the under side partially spotted with cream. The churnings are performed in a room without fire, at a temperature in winter of forty-three to forty-five degrees, and occupy one half to three quarters of an hour.

Several who have adopted my system have reported similar effects — an increase in the quantity with a complete change as to richness of quality. I select from these Mr. John Simpson, a tenant farmer residing at Ripley, in Yorkshire, who, at my request, stated to the committee of the Wharfdale Agricultural Society that he and a neighbor of his, being inconvenienced from a deficient yield of milk, had agreed to try my mode of feeding, and provided themselves with a steaming apparatus. This change of treatment took place in February, 1855. I quote his words:

" In about five days I noticed a great change in my milk; the cows yielded two quarts each, per day, more; but what surprised me most was the change in the quality. Instead of poor winter cream and butter, they assumed the appearance and character of rich summer produce. It only required twenty minutes for churning, instead of two to three hours; there was also a considerable increase in the quantity of butter, of which, however, I did not take any particular notice. My neighbor's cow gave three quarts per day in addition, and her milk was so changed in appearance that the consumers to whom he sold it became quite anxious to know the cause."

My dairy is but six feet wide by fifteen long and twelve high. At one end (to the north) is a trellis window; at the other, an inner door, which opens into the kitchen. There is another door near to this, which opens into the churning-room, having also a northern aspect; both doors are near the south end of the dairy. Along each side, and the north end, two shelves of wood are fixed to the wall, the one fifteen inches above the other; two feet higher is another shelf somewhat narrower, but of like length, which is covered with charcoal, whose properties as a deodorizer are sufficiently established. The lower shelves being two feet three inches wide, the interval or passage between is only one foot six inches. On each tier of shelves is a shallow wooden cistern, lined with thin sheet-lead, having a rim at the edges three inches high. These cisterns incline downwards slightly towards the window, and contain water to the depth of three inches. At the end nearest the kitchen each tier of cisterns is supplied with two taps, one for cold water in summer, the other with hot for winter use. At the end next the north window is a plug or hollow tube, with holes perforated at such an elevation as to take the water before it flows over the cistern.

During the summer the door towards the kitchen is closed, and an additional door is fixed against it, with an interval between well packed with straw; a curtain of stout calico hangs before the trellis window, which is dipped in salt water, and kept wet during the whole day by cold water spirted over it from a gutta-percha tube. On the milk being brought in, it is emptied into bowls. Some time after these bowls (of which a description is given in a former part of this) have been placed on the cistern, the cold-water taps are turned till the water rises through the perforated tube, and flows through a waste pipe into the sewer. The taps are then closed, so as to allow a slight trickling of water, which continues through the day. By these means I reduce the temperature, as compared with that outside the window, by twenty degrees. I am thus enabled to allow the milk to stand till the cream has risen, and keep the skimmed milk sweet, for which I obtain one penny per quart.

Having heard complaints during very hot weather of skimmed milk, which had left my dairy perfectly sweet, being affected so as to curdle in cooking on

being carried into the village, I caused covers of thick
calico (the best of our fabrics for retaining moisture) to
be made ; these are dipped in salt water, and then
drawn over the whole of the tin milk-cans. The con-
trivance is quite successful, and is in great favor with
the consumers. I have not heard a single complaint
since I adopted it.

Finding my butter rather soft in hot weather, I un
covered a draw-well which I had not used since I intro·
duced water-works for the supply of the village and my
own premises. On lowering a thermometer down the
well to a depth of twenty-eight feet, I found it indicated
a temperature of forty-three degrees — that on the surface being seventy degrees. I first let down the butter,
which was somewhat improved, but afterwards the
cream. For this purpose I procured a movable windlass
with a rope of the required length ; the cream-jar is
placed in a basket two feet four inches deep, suspended
on the rope, and let down the evening previous to churning. It is drawn up early next morning, and immediately churned. By this means the churning occupies
about the same time as in winter, and the butter is of
like consistence.

The advantage I derive from this is such that, rather
than be without it, I should prefer sinking a well for
the purpose of reaching a like temperature.

When winter approaches, the open trellis window to
the north is closed, an additional shutter being fixed
outside, and the interval between this and an inner
shutter closely packed with straw, to prevent the access
of air and cold ; the door to the kitchen is at the same
time unclosed to admit warmth. Before the milk is
brought from the cow-house, the dairymaid washes the
bowls well with hot water, the effect of which is to take
off the chill, but not to warm them. The milk is brought
in as milked, and is passed through a sile into the bowls,
which are then placed on the cistern. A thermometer,
with its bulb immersed in the milk, denotes a temperature of about ninety degrees. The hot water is applied
immediately, at a temperature of one hundred degrees

or upwards, and continues to flow for about five minutes, when the supply is exhausted. The bowls being
of thick earthen ware. — a slow conductor, — this does
not heighten the temperature of the milk. The cooling,
however, is thereby retarded, as I find the milk, after
standing four hours, maintains a temperature of sixty
degrees. This application of hot water is renewed at
each milking to the new milk, but not repeated to the
same after it has cooled. The temperature of the dairy
is momentarily increased to above 60°, but speedily
subsides, the average temperature being 52° to 56°.

It will be observed that the churnings in summer and
winter occupy half an hour or upwards. By increasing
the temperature of the cream I could easily churn in
half the time, but I should thereby injure the quality
of the butter. When the butter has come and gathered
into a mass, it is taken, together with the butter-milk,
out of the churn, which is rinsed with water ; the butter is then placed again in the churn with a quantity of
cold spring water, in which salt has been dissolved, at the
rate of one ounce per quart of cream ; after a few minutes' churning, the butter is again taken out; the water
in which it has been washed assumes a whitish appearance. By this process the salt is equally diffused
through the butter, which requires little manipulation,
and is freed from a portion of caseous matter. A recent
analysis of my butter shows only 1.07 instead of 2.45
per cent. of caseine, as before. That it ranks as choice
may be inferred when I state that my purchaser willingly gives me a penny per roll more than the highest
price in Otley market, and complains that I do not supply him with a greater quantity.

In this dairy of the small dimensions I have described,
my produce of butter reaches at times sixty to seventy
pounds per week. Though the size may appear incon-
veniently small, yet I beg to remark on the greater
facility of regulating the temperature of a small in comparison with a large dairy. This difficulty will be found
greater in summer than in winter, as it is far easier to
heighten than depress the temperature.

I have cooked or steamed my food for several years. It will be observed that I blend bean-straw, bran, and malt-combs, as flavoring materials, with oat or other straw and rape-cake ; the effect of steaming is to volatilize the essential oils, in which the flavor resides, and diffuse them through the mess. The odor arising from it resembles that observed from the process of malting: this imparts relish to the mess, and induces the cattle to eat it greedily ; in addition to which, I am disposed to think that it renders the food more easy of digestion and assimilation. I use this process with advantage for fattening, when I am deficient in roots. With the same mixed straw and oat-shells, three to four pounds each of rape-cake, and half a pound of linseed-oil, but without roots, I have fattened more than thirty heifers and cows free from milk, from March up to the early part of May ; their gain has averaged fully fourteen pounds each per week, — a result I could not have looked for from the same materials, if uncooked. This process seems to have the effect of rendering linseed-oil less of a laxative, but cannot drive off any portion of the fattening oils, to volatilize which requires a very high temperature. My experience of the benefits of steaming is such that if I were deprived of it I could not continue to feed with satisfaction.

I have weighed my fattening cattle for a number of years, and my milch cows for more than two years. This practice enables me at once to detect any deficiency in the performance of the animals ; it gives also a stimulus to the feeders, who attend at the weighings, and who are desirous that the cattle intrusted to their care should bear a comparison with their rivals. Another obvious advantage is in avoiding all cavils respecting the weight by my purchasers, who, having satisfied themselves as to the quality of the animal, now ask and obtain the most recent weighing. The usual computation for a well-fed but not over fat beast is, live to dead weight, as 21' to 12, or 100 to 59½, with such modifications as suggest themselves by appearances.

Though many discussions have taken place on the
fattening of cattle, the not less important branch of
dairy treatment has hitherto been comparatively neglected. I therefore venture to call attention to considerations which have arisen from observations in my
own practice affecting the chemistry and physiology,
or, in other words, the science of feeding. That I am
seeking aid from its guidance will be apparent, and I
have no hesitation in admitting that, beyond the satisfaction from the better understanding of my business,
I have latterly derived more benefit or profit from
examination of the chemical composition of materials
of food than from the treatment or feeding experiments
of others which have come under my notice. So persuaded am I of the advantage of this, that I do not feel
satisfied to continue the use of any material, with the
composition of which I am not acquainted, without
resorting to the society's laboratory for an analysis.

*To one leading feature of my practice I attach the
greatest importance — the maintenance of the condition
of my cows giving a large yield of milk.* I am enabled,
by the addition of bean-meal in proportion to the
greater yield of milk, to avert the loss of condition in
those giving sixteen to eighteen quarts per day; whilst
on those giving a less yield, and in health, I invariably
effect an improvement.

When we take into consideration the disposition of a
cow to apply her food rather to her milk than to her
maintenance and improvement, it seems fair to infer
that the milk of a cow gaining flesh will not be deficient
either in caseine or butter.

I have already alluded to the efficiency of bean-meal
in increasing the quantity of butter: I learn, also, from
observant dairymen who milk their own cows and carry
their butter to market, that their baskets are never so
well filled as when their cows feed on green clover.
which, as dry material, is nearly as rich in albumen as
beans. I am also told, by those who have used green
rape-plant, that it produces milk rich in butter. From
this we may infer that albuminous matter is the most

essential element in the food of the milch cow, and that any deficiency in the supply of this will be attended with loss of condition, and a consequent diminution in the quality of her milk.

I am clearly of opinion that you can increase the pro· portion of butter in milk more than that of caseine, or other solid parts. From several, who have adopted my treatment, I learn that on substituting rape-cake for beans they perceive an increased richness in their milk. Mr. T. Garnett, of Clitheroe, who has used bean- meal largely as an auxiliary food for milch cows during the winter season, tells me that when rape-cake is sub- stituted, his dairymaid, without being informed, per- ceives the change from the increased richness of the milk. Mr. Garnett has also used linseed-cake in like quantity; still his dairy people prefer rape-cake.

Mr. Whelon, of Lancaster, who keeps two milch cows for his own use, to which he gave bean-meal and bran as auxiliaries, has recently substituted rape-cake * for bean-meal; he informs me that in a week he saw a change in the richness of milk, with an increase of butter.

The vegetable oils are of two distinct classes: the *drying* or *setting* represented by linseed, the *unctuous* represented by rape-oil. They consist of two proximate elements, margarine and oleine; in all probability they will vary in their proportion of these, but in what degree I have not been able to ascertain. Though the agricultural chemists make no distinction, as far as I am aware, between these two classes of oils, the prac- titioners in medicine use them for distinct purposes. Cod-liver oil has been long used for pulmonary com- plaints; latterly, olive, almond, and rape oils are being employed as substitutes. These are all of the unctuous class of oils. Mr. Rhind, the intelligent medical prac- titioner of this village, called my attention to some experiments by Dr. Leared, published in the *Medical Times*, July 21st, 1855, with oleine alone, freed from

---

* The analysis of cotton-seed cake, in comparison with rape and linseed cake, in a former chapter of this work, will show the comparative value of that as food for milch cows.

margarine, which showed marked superiority in the effect; and I now learn from Mr. Rhind that he is at present using with success the pure oleine, prepared by Messrs. Price & Co., from cocoa-nut oil, one of the unctuous class. That linseed and others of the drying oils are used in medicine for a very different purpose, it seems unnecessary to state.

The oleine of oil is known to be more easy of consumption and more available for respiration than margarine — a property to which its use in medicine may be attributable. If we examine the animal fats, tallow, suet, and other fat, they are almost wholly of the solid class, stearine or margarine, closely resembling or identical with the margarine in plants; whilst butter is composed of oleine and margarine, combining both the proximate elements found in vegetable oils.

It seems worthy of remark that a cow can yield a far greater weight of butter than she can store up in solid fat; numerous instances occur where a cow gives off two pounds of butter per day, or fourteen pounds per week, whilst half that quantity will probably rarely be laid on in fat. If you allow a cow to gain sixteen pounds.per week, and reckon seven for fat, there will only remain nine pounds for flesh, or, deducting the moisture, scarcely three pounds (2.97) per week, equal to .42, or less than half a pound per day, of dry fibrin.

The analyses of butter show a very varying proportion of oleine and margarine fats: summer butter usually contains of oleine sixty and margarine forty per cent., whilst in winter butter these proportions are reversed, being forty of oleine to sixty of margarine. By ordinary treatment the quantity of butter during winter is markedly inferior. The common materials for dairy cows in winter are straw with turnips or mangel, hay alone, or hay with mangel. If we examine these materials, we find them deficient in oil, or in starch, sugar, etc. If a cow consume two stones or twenty-eight pounds of hay a day, which is probably more than she can be induced to eat on an average, it will be equal in dry material to more than one hundred pounds of young grass, which will also satisfy a cow. That one hundred pounds of young grass will yield more butter, will scarcely admit of a doubt. The t venty-eight pounds of hay will be equal in albuminous matter and in oil to the one hundred pounds of grass; but in the element of starch, sugar, etc., there is a marked differ-ence. During the growth of the plant, the starch and sugar are converted into woody fibre, in which form they are scarcely digestible or available for respiration. It seems, then, not improbable that, when a cow is sup-plied with hay only, she will consume some portion of the oleine oil for respiration, and yield a less quantity of butter poorer in oleine.

If you assume summer butter to contain of oleine, . . 60 per cent
"   "   "   "   "   .·   "   of margarine, 40 "   "
                                                            ————
                                                            100 " "·
If the cow consume of the oleine, ... 36 "   "
                                              ————
The quantity of butter will be reduced from 100 to . 64 "   "·

And the proportions will then be, of oleine, . . . 40 ·'   "
"   "   "   "   "   " of margarine, . . 60 "·   "
                                                   ————
                                                   100 "   ·"

If you supply turnips or mangel with hay, the cow will consume less of hay; you thereby substitute a material richer in sugar, etc., and poorer in oil. Each of these materials, in the quantity a cow can consume, is deficient in the supply of albumen necessary to keep up the condition of an animal giving a full yield of milk. To effect this, recourse must be had to artificial or concentrated substances of food, rich in albuminous matter.

It can scarcely be expected, nor is it desirable, that practical farmers should apply themselves to the attain-ment of proficiency in the art of chemical investiga-tions; this is more properly the occupation of the pro-fessor of science. The following simple experiment, however, seems worth mentioning. On several occa-sions, during winter, I procured samples of butter from my next neighbor. On placing these, with a like quan-tity of my own, in juxtaposition before the fire, my

33*

butter melted with far greater rapidity — by no means an unsafe test of a greater proportion of oleine.

The chemical investigation of our natural and other grasses has hitherto scarcely had the attention which it deserves. The most valuable information on this subject is in the paper by Professor Way, on the nutritive and fattening properties of the grasses, in vol. xiv., p. 171, of the Royal Agricultural Society's Journal. These grasses were nearly all analyzed at the flowering time, a stage at which no occupier of grass-land would expect so favorable a result in fattening. We much prefer pastures with young grass not more than a few inches high, sufficient to afford a good bite. With a view to satisfy myself as to the difference of composition of the like grasses at different stages of growth, I sent to Professor Way a specimen of the first crop of hay, cut in the end of June, when the grass was in the early stage of flowering, and one of aftermath, cut towards the close of September, from the same meadow, the analyses of which I give:

HAY, FIRST CROP.		AFTERMATH HAY.	
Moisture,	12.02	Moisture,	11.87
Albuminous matter,	9.24	Oil and fatty matter,	6.84
Oil and fatty matter,	2.68	Albuminous matter,	9.84
Starch, gum, sugar,	39.75	Starch, gum, sugar,	42.25
Woody fibre,	27.41	Woody fibre,	19.77
Mineral matter,	8.90	Mineral matter,	9.43
	100.00		100.00

A comparison between these will show a much greater percentage of woody fibre,—27.41 in the first crop to 19.77 in the aftermath. The most remarkable difference, however, is in the proportion of oil, being 2.68 in the first crop to 6.84 in the aftermath.

On inquiry from an observing tenant of a small dairy farm of mine, who has frequently used aftermath hay, I learn that, as compared with the first crop, he finds it induce a greater yield of milk, but attended with some impoverishment in the condition of the cow, and that he uses it without addition of turnips or other roots, which he gives when using hay of the first crop — an answer quite in accordance with what might be expected from its chemical composition.

It is likewise to be presumed that the quickness of growth will materially affect the composition of grasses, as well as of other vegetables. Your gardener will tell you that if radishes are slow in growth they will be tough and woody ; that asparagus melts in eating, like butter, and salad is crisp when grown quickly. The same effect will, I apprehend, be found in grasses of slow growth: they will contain more of woody fibre, with less of starch or sugar. The quality of butter grown on poor pastures is characterized by greater solidity than on rich feeding pastures. The cows, having to travel over more space, require a greater supply of the elements of respiration, whilst the grasses grown on these poor pastures contain, in all probability, less of these in a *digestible form* available for respiration. The like result seems probable as from common winter treatment — a produce of butter less in quantity, and containing a greater proportion of margarine, and a less of oleine.

It is well known that pastures vary greatly in their butter-producing properties; there is, however, as far as I am aware, no satisfactory explanation of this. If you watch cows on depasture, you observe them select their own food; if you supply cows in stall alike with food, they will also select for themselves. I give rape-cake as a mixture to all, and induce them to eat the requisite quantity; yet some will select the rape-cake first, and eat it up clean, whilst others rather neglect it till towards the close of their meal, and then leave pieces in the trough. Two Alderneys, — the only cows of the kind I have as yet had, — whose butter-producing qualities are well known, are particularly fond of rape-cake, and never leave a morsel. May not these animals be prompted by their instinct to select such food as is best suited to their wants and propensities? If so, it seems of the greatest importance that the dairyman should be informed of the properties of food most suitable for his purpose, especially whilst in a stall, where they have little opportunity of selecting.

It appears worth the attention of our society to make inquiries as to the localities which are known as pro-ducing milk peculiarly rich in butter. When travelling in Germany, I well recollect being treated with pecu-liarly rich milk, cream, and butter, on my tour between Dresden and Toplitz, at the station or resting-place on the chaussée or turnpike-road, before you descend a very steep incline to the valley in which Toplitz is situ-ated. I travelled this way after an interval of several years, when the same treat was again offered. It was given as a rarity, and can only be accounted for by the peculiar adaptation of the herbage of the country for the production of butter.

COMPARISON OF DIFFERENT METHODS OF FEEDING DAIRY COWS.— Being desirous of comparing the result of my method of feeding dairy cows with the system usually practised in this locality, it occurred to me that, as my cows had been accustomed to savory steamed food, a change to ordinary food would be attended with less favorable results than if they had been previously treated in the common mode; and that, under these cir-cumstances, it would be better to institute comparisons with two near neighbors, Mr. Smith and Mr. Pawson, whose practice and results I had the opportunity of inspecting.

Mr. Smith's cow was of rather small frame, but noted for her usefulness as a good milker. At the time of calving her third calf, about the 12th of November, she was in good condition, and gave, soon after, seventeen quarts of milk per day. Her owner states that in the first three weeks (up to the time this comparison was begun) her condition sensibly diminished — a result which I apprehend will be invariable with cows giving this quantity of milk when fed on meadow hay only, with which Mr. Smith's cow was supplied *ad libitum*, and of which she consumed twenty-eight pounds per day. Mr. Pawson's was a nice heifer, three years old at the time of calving her first calf, October 6th, in

more than ordinary condition, and gave about sixteen
quarts per day.   Her owner states that on the first of
January her condition was much diminished.   This is
corroborated by Mr. Myers, a dealer in the village, who
tells me that, previous to her calving, he was desirous
of purchasing her, and would have given from seventeen pounds ten shillings to eighteen pounds, and
describes her as being at that time full of beef.   Her
weight on the first of January, 7 cwt. 2 qrs., bespeaks
her condition as much lowered.

During the month of October, and till late in November, she was turned out in the daytime to graze on
aftermath, and housed during the night, where she was
supplied with turnips.   From the close of November
till the first week in February, her food consisted of

Meadow hay of inferior quality,	. . .	18 lbs. per day.
Swedish turnips,	. . . .	45 " " "
Ground oats,	. . . .	9 " " "

After this the ground oats were discontinued, and
meadow hay of good quality was given *ad libitum*, with
forty-five pounds of turnips.

For comparison I selected a cow of my own, which
calved about the 8th of October, and gave soon after
eighteen quarts of milk per day; she was also of small
size.   At the time of calving her condition was somewhat higher than that of Mr. Smith's.   When the
experiment was begun, on the first of January, no per-
ceivable difference was found in the yield of milk of
Mr. Smith's cow and my own, each giving fifteen and a
half quarts per day.

The following table gives the dates of calving of the
three cows, together with their weights and yield of
milk at the commencement and termination of the
experiment:

When calved.	Yield at calving. Quarts.	January 1.				March 5.			
		Weight.			Yield.	Weight.			Yield
		Cwt.	qrs.	lbs.	Qts.	Cwt.	qrs.	lbs.	Qts.
Mr. Smith's — Nov. 12.	17	8	3	0	15½	8	0	0	9½
Mr. Pawson's — Oct. 6.	16	7	2	0	12	7	1	0	6¼
My own — Oct. 8.	18	9	3	0	15½	10	1	0	12¼

Mr. Smith's cow lost in weight in nine weeks 84 pounds, being 9⅓ pounds per week, with an average yield of 12½ quarts per day. Mr. Pawson's lost 28 pounds. This loss, together with the diminished yield of milk, occurred almost wholly after the oats had been withdrawn; her weight on the 6th of February being still 7 cwt. 2 qrs., and her yield of milk 11 quarts per day.

My cow has gained in the nine weeks 56 lbs., being 6¼ pounds per week, with an average yield of 14 quarts, the diminution being regular. January 1st, 15½; Feb. 4th, 14; March 4th, 12½; making an average yield of 14 quarts per day. The whole loss and gain of weight will be in flesh and fat, the cows having kept up their consumption of food and their bulk.

The weekly account of profit and loss will stand as follows:

	s.	d.
Mr. Smith's cow, average yield for 9 weeks, 12½ quarts per day, at 2d. per quart,	14	7
Deduct loss in flesh, 9½ lbs., at 6d.,	4	8
	9	11
Cost of 14 stones hay, at 6d. per stone,	7	0
Profit,	2	11

		s.	d
Mr. Pawson's cow, average during the first five weeks, 11½ quarts per day, at 2d. per quart,		13	5
Cost of 9 stones inferior hay (at 4d. per stone), per week,	3s. 0d.		
Cost of 63 lbs. ground oats, 4s. 8d.; turnips, 1s. 6d.,	6 2	9	2
Profit,		4	3

		s.	d.
My cow, average yield for 9 weeks, 14 quarts per day, at 2d. per quart,		16	4
Gain of flesh, 6¼ lbs per week, at 6d.,		3	1½
		19	5½
Cost of food :			
Hay, 63 lbs., at 6d. per stone; straw and shells of oats, 1s. 3d.; mangel, 1s.,	4 6½		
Rape-cake, 35 lbs.; bran, 10½ lbs.; malt-combs, 10½ lbs.; bean-meal, 10½ lbs.,	4 0½	8	7
Profit,		10	10½

The richer quality of the manure will probably compensate for the extra labor, cooking, and attention bestowed upon my cow. With a view of extending the comparison, I give particulars of the whole of my cows the weights of which were registered on the 8th of October, and which were still on hand, free from calf, and in a state admitting of comparison. These were bought at a neighboring market in but moderate condition, and were young, having had two or three calves each. A cow in full condition attains her maximum yield in a week or so after calving; whilst those in lower condition continue, by my treatment, to increase their quantity up to about a month after calving.

TABLE.

No.	Calved.	Greatest yield per day.	October 8. Weight.			February 4. Weight.			Yield per day.
		Quarts.	Cwt.	qrs.	lbs.	Cwt.	qrs.	lbs.	Qts.
1.	July 28.	12	9	2	0	10	0	0	8
2.	Aug. 25.	18	10	0	0	11	1	0	14
4.	July 28.	18	8	2	0	10	1	0	15
6.	Sept. 8.	16	10	2	0	10	2	0	14
7.	Sept. 8.	16	10	2	0	11	0	0	10
11.	Aug. 25.	16	9	1	0	9	2	0	11
Average,	. . . .	16	. . . . .			. . . .			12

TABLE — CONTINUED.

No.	March 4. Weight.			Yield per day.	Computed average per day during	Gain, Oct. 8 to Feb. 4.	Gain in weigh per week
	Cwt.	qrs.	lbs.	Quarts.	Weeks. Qts.	Lbs.	Lbs.
1.	10	1	0	8	29 — 10	84	4
2.	11	1	0	14	27 — 16	140	6¼
4.	10	0	0	15	31 — 15	168	8
6.	10	3	0	14	25 — 15	28	1½
7.	11	0	0	10	25 — 13	56	2¾
11.	9	2	0	11	27 — 13½	28	1½
Average,	. . . . . .			12	27½ — 14	. .	. .

My cows, during the period under consideration, were treated as follows: During August and September they were on open pasture by day and housed by night; evening and morning they were supplied with mown grass, and two feeds of steamed mixtuie. Towards the close of September green rape was substituted for the mown grass, with the same allowance of steamed mixture; from the 8th of October, when they were wholly housed, they were supplied with steamed food *ad libitum* three times per day. After each meal ten to twelve pounds of green rape-plant were given, and nine pounds of hay per day till November; from that time steamed food with cabbages or kohl rabi till the early part of February, when mangold wurzel was substituted. It will be observed that I give hay and roots in limited quantities, and the steamed food *ad libitum*. I prefer this to apportioning the cake and other concentrated food in equal quantities to each, as this steamed mixture contains more of the elements essential to milk, and each cow is thus at liberty to satisfy her requirements with it. Nos. 2 and 4, which have given the greatest quantity of milk, have eaten more than their share; whilst No. 1, which has given the least milk, has scarcely eaten more than half the quantity of steamed mixture consumed by 2 or 4. The yield of milk and the live weights on the 4th of February and the 4th of March scarcely vary. During February thirty-four pounds of mangold were substituted for kohl rabi; with this change the cows became more relaxed. My experience in weighing, extending over several years, has shown me that when animals, from change of food, become more relaxed or more costive, their weighings in the former state denote less, whilst in the latter they denote more, than their actual gain in condition. I have known instances in which a month's weighing, accompanied by relaxation, has shown no gain, whilst, with restored consistency, the gain doubled.

I now proceed to examine the materials of food, their composition, and the probable changes they undergo in the animal economy.

Quantity and description of food supplied to six cows during twenty-seven and a third weeks, and its composition in proximate elements and minerals.

	Per day.	Total weight of food given.	Cost per ton.			Total cost.			Weight of food when dried.
	lbs.	lbs.	£	s.	d.	£	s.	d.	lbs.
Meadow hay,	56	10,715	4	0	0	19	2	9	9,420
Rape-cake, .	30	5,740	6	10	0	16	12	t	5,456
Malt-combs,	9	1,722	5	9	0	4	3	0	1,660
Bran, . . .	9	1,722	6	10	0	5	0	0	1,500
Beans, . . .	9	1,722	9	6	8	7	3	6	1,500
Green food,	204	39,032	0	10	0	8	14	6	5,740
Oat-straw, .	50	9,566	1	15	0	7	9	0	8,407
Bean-straw,	12	2,296	1	15	0	7	16	0	1,964
Total, . .	379	72,515				70	0	9	35,647

	Albumen.	Starch.	Oil.	Fibre.	Minerals.
	lbs.	lbs.	lbs.	lbs.	lbs.
Meadow hay,	990	4,257	287	2,933	953
Rape-cake,	1,803	2,177	611	494	171
Malt-combs,	411	791	51	320	88
Bran, . . .	246	800	96	258	100
Beans, . . .	464	774	34	176	53
Green food,	862	3,074	115	1,148	541
Oat-straw, .	287	3,066	100	4,526	428
Bean-straw,	376	725	51	594	217
Total, . .	5,439 = Nitrogen 888 lbs.	15,664	1,345	10,449	2,551

ANALYSIS OF MILK BY HAIDLEN

Water, ...	873.
Butter, ...	30.
Caseine, ...	48.2
Milk sugar, ...	43.9
Phosphate of lime, ...	2.31
Magnesia, ...	.42
Iron, ...	.07
Chloride of potassium, ...	1.44
Sodium and Soda, ...	.66
	1000.00

34

Production of milk by 6 cows, average 14 quarts per day each, for 27½ weeks = 16,072 quarts, which at 41 oz. per quart = 41,184 lbs.

lbs.

When dry or free from moisture, ...   5230

Butter in 16,072 quarts, at 30 per 1000	=	1235
Caseine in " " " 48.2 per 1000, ...	=	1977
Sugar of milk, ...	=	1804
Minerals. {Phosphate of lime, ... 99 }	=	214
{Other, ... 115 }		

5230

Gain of weight 500 lbs., of which I compute 300 lbs. as fat.

200 lbs. as flesh.

500

Nitrogen, ... 316 lbs.

Phosphate of lime, ... 99

Phosphoric acid, ... = 45.50

Cost of food per cow per week,   . 8s. 6½d.

When the yield of milk is less, the cost of food is reduced to 7s. 8d per week.

	s.	d.
Gross return in milk, ...	16	4
" " " weight, ...	1	6
" " " manure, ...	2	8
	20	6

ANALYSIS OF EXCREMENT BY PROFESSOR WAY.

Per cent.

Moisture, ...	84.85
Phosphoric acid, ...	.39
Potash, ...	.58
Soda, ...	.22
Other substances, ...	13.96

100.

Nitrogen, ...41

Ammonia, ...49

Manure, 88 lbs. per cow per day.

For 6 cows per day 528 lbs. = 3696 lbs. per week.

" " " for 27½ weeks 101,028 lbs., containing of

Nitrogen, ...	414 lbs
Phosphoric acid, ...	393
Potash, ...	585

```
Nitrogen incorporated in food, . . . . . . . .   888 lbs
        Caseine, . . . . . . . . . .   316.
        Fibrin, . . . . . . . . . .     7.35
        Manure, . . . . . . . . .     414.
Balance consumed in perspiration, . . . .   150.65
                                           ───────
                                           888.00
```

The materials of food are shown to have cost . . £70 0s 9d.

	£.	s	d.
Gross value 16,072 quarts of milk, at 2d. per quart, .	133	18	8
Gain of weight 500 lbs., at 6d. per lb., ...	12	10	0

	£.	s.	d.			
Nitrogen in manure 414 lbs. = Ammonia 494 lbs., at 6d., ...	12	7	0			
Phosphoric acid 393 lbs., at 1½d. per lb., .	2	9	1			
Potash 585 lbs., at 3d. per lb., ...	7	6	3			
				22	2	4

```
                                           £168 11 0
```

Manure per cow per day 88 lbs., per week 616 lbs.	s.	d.
Containing ammonia 3 lbs., . . .	1	6
Phosphoric acid 2.40 lbs..	0	3½
Potash 3.57 lbs., ...	0	10½

Value of a cow's excrement, per week, ... 2 8

The analyses of the chief ingredients of my own produce, or such extra materials as I usually purchase, have been made by Professor Way ; for other materials I have had recourse to a very useful compilation by Mr. Hemming (vol. xiii., p. 449, of the Society's Journal), and to Morton's "Cyclopædia of Agriculture." The analysis of straw is that of oat-straw ; that of green food is derived from the analysis of rape-plant, cabbages, and kohl rabi. During February and March I have been using wheat and barley straw with mangold, and, as those materials contain less oil, I give in the steamed food three ounces of linseed-oil per day to each animal. For the composition of milk I adopt that by Haidlen, whose method of analysis is reputed to be the most accurate, the proportion of butter in my milk being this season very similar to that given by him.

It will be observed that this is the gross return for twenty-seven and one third weeks from the time of calving, from which will have to be deducted expense of attendance, etc.

	£.	s.	d.
The materials used for food are found to have cost . .	70	0	9
The value of these materials as manure consists of 888 lbs. nitrogen = 1061 lbs. ammonia, at 6d.,. . . .	26	10	6
Phosphoric acid and potash, ... :	9	15	4
Value of food if employed as manure, . . . £36		5	10

The 16,072 quarts of milk, at 2d. per quart for new milk, at which price it enters largely into consumption as food for man, amount to £133 18 8

The nitrogen in the milk 316 lbs. = ammonia	£.	s.	d.
378 lbs., at 6d. per lb., ...	9	9	0
Phosphoric acid in ditto. 45½ lbs., at 1½d. per lb.,	0	5	8
		£9	14	8

From these statements it will be seen that materials used as *food for cattle* represent double the value they would do if used for manure, whilst that portion converted into *food fitted for the use of man* represents a value thirteen to fourteen times greater than it would as manure.

It then appears clear that it is for the feeder's profit to use his produce as much as possible as food for cattle, with the view to convert it with the utmost economy into food for man, and *thus increase rather than enrich his manure-heap.*

The calculation of caseine in milk is based upon the supposition that my milk is equal in its proportion of that element to that analyzed by Haidlen. Several analyses by other chemists show a less percentage, 4 to 4.50. As my cows are adequately supplied with albuminous matter, I have a right to presume on their milk being rich in caseine.

The loss of nitrogen by perspiration, 150.65 lbs., is nearly 17 per cent. Boussingault found a loss of 13.50 of nitrogen in a cow giving milk.

The abstraction of nitrogen in the milk is computed at	£.	s.	d.
316 lbs., value, ...	9	9	0
The abstraction of phosphoric acid in the milk is computed at 48½ lbs., ...	0	5	8

Either the rape-cake or bran alone suffices for the restoration of the phosphoric acid.

The amount of phosphoric acid in the manure is 393 pounds, being about sixteen per cent. of the whole ash or mineral matter. The ash of meadow hay contains about 14 per cent., that of rape-cake 30 per cent., bran 50 per cent., malt-combs 25 per cent., and turnips, &c., 10 per cent. of phosphoric acid.

The amount of potash in the excrement is 616 pounds, being about 25 per cent. of the whole ash or mineral matter. The ash of meadow hay contains about 20 per cent.; rape-cake, 21 per cent.; malt-combs, 37 per cent.; turnips (various), 44 per cent.; from which it may be inferred that the sample of excrement sent to Professor Way for analysis did not contain more than a fair proportion of these ingredients.

To ascertain the quantity of excrement, the contents of the tanks into which the cows had dropped their solid and liquid excrement during five weeks were weighed, and found to be 500 cwt. 2 qrs. 0 lbs , from 18 cows, being 88 lbs. per cow per day. The sample for analysis was taken from that which the cows had deposited within the preceding 24 hours. This was collected in the mud-cart, well blended, and sent off quite fresh.

It is sufficiently proved, by the experience of this district, that 20 pounds of meadow hay suffice for the maintenance of a cow of fair size in store condition; a like result is stated to be obtained from 120 pounds of turnips per day. The six cows will have then required. during the 27½ weeks, for their maintenance, only

Per day.		Weeks.	Total Weight.		Albumi-nous mat.	Oil.	Starch. &c.
lbs.			lbs.				
120	of hay or for	27½	22,960	containing of	2127	616	9130
150	of turnips, or for	27½	137,760	" "	2295	306	9100

They will further have required adequate food —

34*'   26

	Albuminous matter, fibrin, and caseine.	Oil and butter.	Starch and sugar of milk.
For the production of	2,116	1,235	1,894
And for maintenance by turnips, . .	2,295	306	9,100
	4,411	1,541	10,994
The food supplied is computed do have contained	5,459	1,345	15,664

I omit the minerals, which are observed to be in excess of the requirements.

For the maintenance of a fair-sized cow, for one day, in a normal state, the following elements seem adequate:

	Albumen.	Oil.	Starch, &c.	Lime.	Mineral ingredients. Phosphoric acid.
In 20 lbs. of hay, .	1.85	.536	7.95	.90	1.11
In 120 " " turnips,	1.98	.26	7.82	.97	1.9

When cows are in milk, there occurs a much greater activity of the functions; they eat and drink more, evacuate more excrement, and, in all probability, spend considerably more food in respiration. Whilst the 17.60 lbs. per day dry matter in 20 lbs. of hay are found adequate for the maintenance of a cow in a store state, the six cows in milk have eaten on the average 21.37 lbs. solid matter per day during the 27½ weeks. When I have fattened cattle together with a number of milch cows of similar size, which gave on an average eight quarts of milk per day, the whole being fed with moist steamed food, and receiving the same allowance of green food, I have found the fattening cattle refuse water, whilst the milch cows on the average drank upwards of 40 pounds per day of water given separately. The eight quarts of milk contain only about 17.58 lbs. of water; still, in several analyses of excrement, I have noticed little difference in the percentage of moisture in that from the fattening animals as compared with that from cows giving milk.

These facts would seem to show that upwards of 20 lbs. more water were given off from the lungs and pores of the skin of a milking than of a fattening animal.

The excrement of the six milch cows, 88 lbs. per day on the average, is found to contain of nitrogen 36, equal to that in 2.25 lbs. of albumen; whilst 1.85 of albumen in the 20 lbs. of hay is found adequate for maintenance.

On comparing the supply of the food to the six milch cows with their requirements and production, there seems an excess in the albuminous matter, a deficiency in the oil for the fat and butter, an excess in the starch, &c. Taking, however, the increased activity of the animal functions, and consequent consumption of food by the milch cow, I am not encouraged to lower my standard of food. That it has sufficed is abundantly proved, as each of the six cows under observation has gained in condition during 27½ weeks.

My observations on nutrition tend to the conclusion that if you supply animals with starch, sugar, &c., to satisfy their requirements for respiration, you enable them to convert the oil of their food into butter or fat to such extent as their particular organism is fitted for effecting it.

On the 12th of March I purchased Mr. Smith's cow (see p. 392) for twelve pounds ten shillings, being more than her market value, for the purpose of trying her on my food; her yield of milk had then diminished to 8 quarts per day. On the 31st of March, four weeks from the former weighing, and nineteen days after being treated with my food, her yield of milk had increased to 9½ quarts per day, and her weight to 8 cwt. 1 qr., being 28 lbs. increase.

Mr. Pawson's cow, which was continued on the same food, namely, meadow hay *ad libitum*, and a more limited supply of turnips, reduced her yield of milk to less than 5 quarts per day, without alteration in her weight.

My cow first placed on trial with those of Mr. Smith and Mr. Pawson gave a yield of milk of 12 quarts per day, and gained 28 lbs. in the four weeks, her weight on the 31st of March being 10 cwt. 2 qrs.

The weight and the yield of milk of the six, on the 31st of March, were:

	March 4.			Yield of milk per day.	March 31.			Yield of milk per day.	Gain in 4 weeks.
	cwt.	qr.	lbs.	quarts.	cwt.	qr.	lbs.	quarts.	lbs.
Weight of No. 1.	10	0	26	8	10	3	0	8.9	58
" " " 2.	11	1	0	14	11	3	0	14.9	56
" " " 4.	10	0	0	14½	10	1	0	13	28
" " " 6.	10	3	0	14	11	2	0	12	84
" " " 7.	11	0	0	10	11	3	0	10	84
" " " 11.	9	2	0	11	10	1	0	12	84

On referring to the previous weighing, there was little or no gain from Feb. 4th to March 4th, the cows being at that time in a somewhat more relaxed state. During March they wholly regained their consistency. The gain shown in the weighing, March 31, by the six cows, appears therefore unusually great. It should, however, be computed as made during the eight weeks from Feb. 4th to March 31, being with an average yield of nearly 12 quarts (11.66) per day each, at the rate of 8⅙ lbs. each per week on the average.

No. 11, it will be observed, is stated as giving more milk on the 31st than on the 4th of March. It occasionally happens that cows drop their yield of milk· for a day or two, and then regain it, especially when in use. The whole of these six cows were kept free from calf till February, when Nos. 2 and 4 were sent to bull. I had some hesitation in regard to No. 4, from her having suffered from pleuro. Her milk, tested by a lac-tometer, denoted a less than average proportion of cream ; still, in quantity, and keeping up its yield for a length of time, being of more than ordinary capability, I decided to retain her.

Nos. 1 and 7, which are giving respectively 8 and 10 quarts per day, are in a state of fatness ; they will probably be sold in June as prime fat, when their yield of milk will probably be 6 and 8 quarts per day each.

They may be expected to fetch twenty pounds to twenty-three pounds. No. 6 is also in a state of forwardness. No. 11, which suffered considerably from pleuro, is in comparatively lower condition. During the season from the close of October to the close of January, I avoid purchasing near-calving cows, which are then unusually dear, my replenishments being made with cows giving a low range of milk, and intended for fattening. I find them more profitable than those which are quite dry. The present season I had additional grounds for abstaining from buying high-priced cows, from the recent presence of pleuro.

On the 2d of March I had occasion to purchase a calving cow, which was reported to have calved on the 28th of February. Her weight on the 4th of March was 9 cwt. 1 qr. I supplied her with 35 lbs. of mangold, and hay *ad libitum*, of which she ate 22 lbs. per day. The greatest yield she attained was somewhat more than 13 quarts per day. On the 31st of March her weight was 9 cwt., being a loss of 28 lbs. in four weeks. Her yield of milk had diminished to 11¼ quarts per day. A week after this her milk, during six days, was kept apart, and averaged 10 quarts per day ; being at first rather more, at the close rather less, than this. The cream produced from these 60 quarts was 9 pints, the butter 63 oz. The butter from each quart of cream was 14 oz. The proportion of butter to milk was 63 oz. from 60 quarts — rather more than 1 oz. per quart.

An equal quantity of milk from a cow (calved Oct. 8th) treated with steamed food, and set apart for comparison, gave less than 7 pints of cream, which produced 79 oz. of butter.

In quality and agreeableness the butter from steamed food and cake was decidedly superior to that from hay and mangold.

Mr. Stansfeld, of Chertsey, has supplied me with the following interesting particulars of two Alderney cows which were treated as follows :

From Dec. 1st to-Jan 15th, with Swedes and meadow hay.

From Jan. 15th to Feb. 17th, pulped and fermented Swedes, meadow hay, and 3 lbs. rape-cake, 2 lbs. bean-meal, 2 lbs. bran, 2 lbs. malt-combs.

From Feb. 17th to May 1st, 5 lbs. rape-cake, 2 lbs. bran, 2 lbs. malt-combs.

Results :

December 1st to January 15th, yield of butter from each quart of cream, $10\frac{3}{4}$ oz.

January 15th to February 17th, yield of butter from each quart of cream, 14 oz.

February 17th to May 1st, yield of butter from each quart of cream, $18\frac{2}{3}$ oz.

The yield of butter in proportion to milk, Dec. 1st to Jan. 15th, is described as unsatisfactory.

The yield of butter in proportion to milk, Feb. 17th to May, as 2 oz. per quart, which is their maximum proportion.

Soon after calving the two cows gave 18 quarts of milk per day; on the 15th of May, 15 quarts per day.

Mr. Stansfeld has completely satisfied himself that by the process of fermentation the turnip loses its disagreeable taste, and that his butter is of excellent quality.

If I take the supply of turnips, 120 lbs. per day, as requisite for the maintenance only of the cow, the nutritive elements will be :

Albumen.	Oil.	Starch and sugar.
1.79	.264	7.92

Reckoning the oil as used for respiration, and computing it
    in proportion of 5 to 2 as compared with starch = . .66

                                                  8.58

The food supplied to the cow consists of :

	Lbs.	Water.	Dry.	Albu-men.	Oil.	Starch and sugar.	Fibre.	Mine-rals.	Phosphoric acid.
Hay, . . .	22	$2\frac{1}{4}$	19.36	2.03	.59	8.74	6.05	1.95	.30
Stored mangold,	35	28.0	7.	1.05	..	4.20	1.05	.70	.05
			26.36	3.08	.59	12.94	7.10	2.65	.35

oz.

The 13 quarts of milk yielded of butter, . . ' . . 13.60
Deduct for moisture, &c., ... 2.28

11.32
Butter in the skimmed milk estimated as 68

12.00 oz
12 ounces of pure oil in the butter are ¾ lb. = . . .75

lb.
The oil in the food, ...59
The starch and sugar, ... 12.94
Used for animal respiration, ... 8.58
—— 4.36

There appears, then, in this supply of food, .59 lbs. oil and 4.36 lbs. starch for the production of .75 in the butter from 13 quarts per day, the cow's greatest yield. At the time the milk was tested, aftermath hay was sub-stituted for first-crop hay, in equal quantity. This, it will be observed, is decidedly richer in oil. Her prod uce· had lessened to 10 quarts per day ; her production of butter was 10.50 oz. per day, or of pure oil about 9 oz. ; for the supply of oil the aftermath hay alone would be much more than adequate.

On examining the adequacy of the food for the sup-ply of albumen for the caseine,

lbs.
I find this to be, ... 3.08
I assume that in 120 lbs. of turnips, as required for
    maintenance, in a normal state, ... 1.98

1.10

Which, according to Haidlen's analysis, will be adequate to the supply of 8.60 quarts per day. The supply of mineral substances is in excess.

The cow, under this treatment, gave,

Soon after calving, fully . . . . . 13 quarts per day.
Five weeks after calving, ... 11¼ " " "
In less than 8 weeks after calving, . . 9 " " "

And with this there occurred also a loss of weight.

We find this cow supplied with food amply rich in

every element suited to her wants and purposes, with
the exception of the nitrogenous principle only, lowering her condition, and likewise her yield of milk, till it
approaches a quantity for which her food enables her
to supply a due proportion of caseine.

About the 20th of April, the cow's yield being reduced to 9 quarts per day, her food was changed to
steamed mixture. Soon after this her yield increased
to 11 quarts per day. Her weight, April 28th, 9 cwt.;
May 16th, 9 cwt. 14 lbs.: yield of milk, 11 quarts.

I now introduce the dairy statistics of Mr. Alcock, of
Aireville, Skipton, who has for some time been prac-
tising my method of treatment, with such modifications
as are suited to his circumstances.

During the winter season, Mr. Alcock's food consisted
of mangold, of which he gave 20 lbs. per day to each,
uncooked, together with steamed food *ad libitum*, consisting of wheat and bean straw, and shells of oats.

Carob bean and Indian meal, for each, .	. 3 lbs.	per	day.
Bran and malt-combs, ...	. 1 "	"	"
Bean-meal, ...	. 3½ "	"	"
Rape-cake,*	. 3 "	"	"
Of extra food, ...	11¼		

From March 19, when his store of mangold was exhausted, he increased his supply of Indian meal to 4
lbs. per day, and omitted the carob bean.

During the month of January, Mr. Alcock obtained
from 759 quarts of milk 1323 oz. of butter, being from
each 16 quarts 26⅝ oz.; during February and March,
from 7368 quarts of milk 12,453 oz. of butter, or from
each 16 quarts fully 27 oz.: so that rather less than
9½ quarts of milk have produced 16 oz. of butter.
The average produce from each quart of cream was
20½ oz.

Mr. Alcock fattens his cows whilst giving milk, and
sells them whilst giving 4 to 6 quarts per day. He

---

* The rape-cake used by Mr. Alcock was of foreign manufacture, evidently rich in oil, but containing mustard, and on this account supplied in
less proportion.

quite agrees with me that it is far more profitable to buy far-milked cows for fattening; and obtains, from a change to his food, 2 to 3 quarts per day more than the cow had given previously.

Though Mr. Alcock's cream is not so rich as what I have described on pp. 377 and 378, it is more than ordinarily so. His mode of separating his milk from his cream differs from my own, his milk being set up in leaden vessels, from which, on the cream being formed, the old milk is drawn, by taking a plug from a hollow tube, with perforated holes in the centre of the vessel. To this difference I am disposed in some degree to attribute the less richness of Mr. Alcock's cream. On examining the cream with a spoon, after the dairy-keeper had drawn off the milk, I observed some portion of milk, which would have escaped through my perforated skimmer.

Mr. Alcock's proportion of butter from milk, which is the matter of practical importance, is greater than what I have shown on a preceding page, being from each 16 quarts of milk 27 oz. of butter.

QUALITY OF BUTTER. — In January, 1857, samples of about 56 oz. each, of butter of my own, and also of Mr. Alcock's, were sent to the laboratory of Messrs. Price & Co.'s candle-works, at Belmont.

My butter was found to consist of (taking the pure fat only),

Hard fat, mostly margarine, fusible at 950°, . . 45.9
Liquid, or oleine, ... 54.1
—————
100.0

Mr. Alcock's,

Hard fat, mostly margarine, fusible at 10°, . . . 36.0
Liquid, or oleine, ... 64.0
—————
100.0

For these analyses of butter the agricultural public is indebted to the good offices of Mr. George Wilson, director of Messrs. Price & Co.'s manufactory. It will be observed that Mr. Alcock's milk is richer in butter, and that his butter is also richer in proportion of oleine to margarine than my own.

Professor Thompson ("Elements of Agricultural Chemistry," 6th edition, p. 317) states that winter butter consists more of solid, and summer more of liquid or oleine fat.

An analysis of butter made in Vosges gives:

	Summer.	Winter.
Solid or margarine fat,	40	65
Liquid (or oleine) fat,	60	35
	100	100

In Lehmann's "Physiological Chemistry" (Leipsic edition, vol. ii., p. 329), an analysis of butter by Bromus gives:

Margarine,	68
Oleine,	30
Special butter-oil,	2
	100

It will be observed that my butter may be classed as summer butter, and that Mr. Alcock's is the richest in the proportion of oleine. Both were produced in the month of January.

These results are important, and completely establish the conclusion I had previously formed, that the quantity and quality of butter depend essentially on the food and treatment; and that by suitable means you can produce *as much and as rich butter in winter as in summer*.

PLEURO-PNEUMONIA.

In the chapter on the Diseases of Dairy Stock, p. 271, allusion only was made to pleuro-pneumonia as one of the fatal epizoötics that have from time to time decimated the cattle of Europe. At the time the first editions of this work appeared, no instances of this terrible scourge had, to my knowledge, appeared in this country.

During the year 1859, however, several cases occurred in Massachusetts and New Jersey, which, from their symptoms both before and after death, can leave little or no doubt of their being genuine pleuro-pneumonia, while at the same time they add weight to the already conclusive testimony that the disease is contagious or infectious in its character. Whatever modification may appear in the symptoms exhibited in the cases in this country, as compared with those in England and on the continent, may be readily accounted for on the ground of difference of climate, treatment, &c.

This dangerous and fatal disease derives its name from the parts affected. The pleura is the membrane which covers the lungs and lines the cavity of the chest, and pneumonia the substance of the lung itself. Pleuro-pneumonia is applied to the compound disease in which both these parts are attacked, and which, in its early stages, appears to be of an inflammatory character. The lungs are found, on a post-mortem examination, to have lost their light, porous consistence, and their pinkish color, and to have become very dark, condensed, or consolidated, filled with lymph to such an extent as to be impervious to air and incapable of expansion and contraction, indicating, of course, that they had lost the power of vitalizing the blood, when the animal must die. A large body

of water is often found in the chest, as is observed in
cases of pleurisy.

The early symptoms of pleuro-pneumonia are often
quite obscure, and would not be perceived where the disease was not suspected, and the animal carefully watched,
and perhaps not even then till it had considerably advanced. The interior of the eyelids becomes red, while
in the healthy animal it is a beautiful rose color ; the
pulse increases five or six beats over its usual activity,
that of the healthy animal, from five to eight years, being
about forty-eight or fifty a minute, that of the young animal being quicker — sometimes even as high as sixty.
The respirations are increased in activity from five to
ten per minute, the natural activity being about seventeen per minute. The noise made in breathing, as the
ear is placed upon the chest or just behind the elbow, becomes louder, and resembles somewhat the crumpling of
paper. If the sides are struck, the animal suffers more
than usual, and there appears, morning and evening, a
slight, dry cough, often short and painful. This is the
first stage of the malady, and would not attract attention,
since the animal may still continue to eat, drink, ruminate, labor, give milk, &c., apparently as usual. In this
stage it is curable under careful treatment.

Then the trouble rapidly increases. The appetite diminishes ; there is a disinclination to chew the cud, and it
is done by jerks ; the hair is dull and staring ; the temperature of the skin and external surfaces is very uneven ; the
horns may be cold and warm alternately, or the legs may
appear very cold, and the horns or other parts of the
body hot. If in pasture, the animal withdraws from the
rest of the herd ; in four or five days after the disease is
seated, the appetite ceases entirely ; the breathing becomes quicker and more labored, the respirations increasing to thirty, forty, or even forty-five per minute ; the
nostrils are somewhat dilated, discharging a light, mucous
substance ; the animal lows, and appears to suffer ; in
some cases it swells up. The cow falls off in milk. In
pressing even lightly upon the back, just behind the
withers, the animal shows great pain. The breath grows

warmer, and often fetid ; the danger rapidly increases, of course. The animal will often press her muzzle very hard against the partition as if for support, and breathe from the mouth, catching her breath with difficulty, and soon dies. The progressive symptoms vary greatly, however, in different animals ; but the cough is the key note of the disease, and appears in all.

It is only in the early stage of the disease that it is curable ; and even if apparently cured, it is probable that the relief is only temporary, and that the disease is latent in the system, and ready to appear with renewed force on the occurrence of any exciting cause. After the very early stages, therefore, it is best to kill and bury the animal, and thus save cost and risk of infection.

There seems to be no longer room for doubt that the disease is contagious or infectious. It appears to be communicated by animal poison in the air proceeding from the lungs and breath, or the respiratory surfaces of a diseased animal ; and any animal of the same species, coming in contact or within the influence of this vitiated air, is very liable to be infected. It attacks old animals and young, cows in milk or otherwise, calves and oxen, indiscriminately.

From Collot, the author of a recent and valuable French work on the dairy cow, (*Traité spécial de la Vache laitière,*) who speaks of this disease, I translate as follows : " This malady is the greatest scourge which could fall upon the farmer ; it is hereditary and contagious, and hence it will rarely disappear, or rather never disappear, from a country which it has once invaded. To my mind, the terrible typhus is less to be dreaded than pleuro-pneumonia, because if it strikes severely it may disappear, and is not persistent ; the *evil* is only temporary ; while with pleuropneumonia it is lasting, contagious and endemic, or latent, and ready to break out on any exciting cause. It is then the most terrible of maladies which could threaten our most valuable herds of cattle ; and I cannot comprehend the apathy of the government with regard to so great a calamity, which is insensibly extending in France, and endangering the most powerful lever of our agriculture, neat

35 *

cattle, — the most important production, and that which ought most to be encouraged, that of beef. The German countries give us an example of energetic measures. Why should we hesitate to follow them ?

"When the invasion is well ascertained, public functionaries should advise the destruction of all the cattle in the barn where the disease has established itself. If the owner refuses to take this advice, good as well for him as for the public at large, the public officer ought to do all in his power to hem in the disease, and to prevent the animals from an infected barn from being brought in contact with others in the pastures, or to be driven to the markets and the fairs. In fine, it will be necessary to establish around the locality of the infection a kind of *cordon sanitaire*, to notify the prefect and the minister of agriculture, and to raise a loud cry of alarm, because no malady has ever done so much evil as pleuro-pneumonia."

The outbreak of this disease can be traced invariably to the introduction of cattle from abroad, and its spread and extension can only be prevented by the immediate and complete isolation of the infected animals from others, or the destruction of all animals in which premonitory symptoms appear, and those which have been exposed to the infection.

As already intimated, the first stage of the disease is the only period when it can be cured ; and after it has become fixed upon the lungs, dosing is of little use, and the animal ought to be destroyed.

In the first stage, Collot recommends " bleeding slightly in the neck, and rubbing the whole body for half an hour with whisks of straw, and then to cover the animal and leave her alone. Three or four hours after bleeding he would give an emetic in warm water, followed by eight similar doses two hours apart ; during the intervals of the two hours, moderate quantities of the following beverage : —

"Boil two or three quarts of barley for ten minutes in about two gallons of water ; then pour off this water, which contains the acrid principles of the grain, and replace it by about five gallons of fresh. Boil this an hour, and let it cool till lukewarm ; then add two pounds of sulphate of soda or Glauber's salts. Administer doses of this water, strained through a linen cloth, four times a day. Continue this treatment three, four, or five days, until the animal is better. A second bleeding at the neck, if it can be done, if not, from the large vein in the belly, may take place eight or ten hours after the first.

"When the animal is better, give it at first some clear, warm water, and soon after increase its ration of hay, fresh grass or roots cut and mixed with barley meal, and a moderate dose of table salt. The temperature of this water may be gradually diminished, till in a few days the animal returns to its usual condition. As a diet, during treatment, oatmeal is undoubtedly one of the best articles ; and it may be made into a thin gruel, with salt enough to make it palatable.

"If during the preceding treatment the animal should cough a little, and respiration be quick and labored, with an apparent pain in the chest, the tender parts should be rubbed with the following preparation : —

$\frac{1}{2}$ oz. pulv. cantharides, (Spanish flies.)
$\frac{1}{4}$ " euphorbia, (a powerful irritant.)
1 pint of alcohol.

Mix in a small earthen jug, put the cork in loosely, and warm and shake it up, then pass through a linen strainer, and preserve it for use as a counter-irritant on the sides of the chest. Rub the tender parts of the chest in order to produce irritation, which will terminate in small blisters containing a reddish liquid. Some have used successfully a common mustard seed poultice placed on the sides of the chest, after shaving off the hair from the parts ; but the above preparation of Spanish flies is preferable.

"If the animal coughs frequently, and the discharge from the nostrils is thick and yellow, and there is a rattle in the air passages, prepare the following fumigation : —

" Boil two handfuls of mallows in water for half an hour, and place it, while boiling, beneath the nose of the animal, having enveloped its head with a cloth, so that it is obliged to breathe the vapor. Repeat this fumigation four or five days. If this discharge continues, pass a seton through the dewlap, using with it the root of black hellebore boiled half an hour in vinegar.

"The following may be made use of instead of the above : —

1 oz. sulphate of alumina or potassa.
1 "   sulphate of zinc.
1 "   Spanish powders.
1 "   oil of turpentine.
⅓ "   camphor.

Reduce these to powder, dissolve in one quart of strong vinegar, mix in a bottle, and shake it well. Raise the head of the animal, and turn a small spoonful into the nasal passages. The animal will sneeze powerfully, and throw out the thick mucus which obstructs the air passages. Repeat this practice for several days.

"If the disease resists this treatment, and the animal refuses to eat or ruminate, or if, after having eaten, the belly is swollen, the animal froths at the mouth, lows frequently, and is unable to lie down, it is better to kill it at once, and not, while losing time, add to the danger of contagion.

"Pleuro-pneumonia has not hitherto attacked any but neat cattle ; it has not extended to horses, among which the contagion is not to be apprehended."

Mr. Winthrop W. Chenery, of Belmont, Mass., who has lost a large number of valuable animals by this malady, wrote to his correspondents in Holland for information in regard to the existence of the disease in the locality from which some of his cattle were obtained, and the modes of treatment recommended by distinguished veterinary surgeons there, and received the following reply, which he has very kindly placed in my possession : —

"There was no disease prevailing at the stables where the cows were procured, although a disease is existing throughout the whole country, (Holland,) known as 'phthisis'—a pulmonary disease. The governments of France and Holland have offered large sums to whoever shall discover a remedy ; yet none has as yet been found.

Cattle infected with this disease suffer a long time before it is observable ; and when first noticed, they are usually sold to the butcher, in order to be killed for food.

" There is, however, much benefit to be derived from inoculating the healthy animals. This inoculation is done near the end of the tail. The hair is clipped off, the skin cleaned, and two incisions made with a lancet, into which the virus is introduced. The virus must be obtained from the lungs of a cow suffering with the disease, and killed for the purpose, and not from an animal that has died in the natural way from the effects of the disease. The manner of obtaining it is to cut off a portion of the lung between the healthy and the infected parts, the part marbled like water. and the blood is wrung out into a vessel and allowed to stand one day, when the bloody part will sink to the bottom, and a lemon-colored liquid will remain upon the surface. This, *if free from scent*, is fit for use, and may be preserved in a vial. In cold weather it will keep eight or ten days before becoming too corrupt for use, while in warm weather it will hold good only one or two days.

" The drops introduced into each incision will produce, in a week or fortnight, and in some cases a longer time, a pock quite similar to that caused by the inoculation of persons with the cow pox. When no pock appears, it is presumed that the animal is not susceptible to the disease. When the tail of the animal becomes much swollen, an incision is made, in order that the infectious matter may run out, and the wound is from time to time cleansed with water.

" The benefits resulting from this discovery are such that where the peasants formerly lost from fifty to sixty per cent. of their cattle, they now lose only one per cent.

" Inoculation is also practised on animals afflicted with the disease, and sometimes with favorable results. Some have resorted to bleeding, some have purged with English salt and water, others have fumigated and purified their stables, but no sufficient remedy has been found."

There is, it is proper to say, a difference of opinion among scientific practitioners in regard to the efficacy of inoculation—some contending that it will produce the identical disease, and infect the animal as injuriously as if taken from the breath of a diseased animal, and others maintaining that the preponderance of the testimony is strongly in its favor. The reports of experiments of the Dutch, Belgian, and other commissions appointed to investigate this particular point, are not very conclusive, though the results of the most extensive series of experiments appear very strongly to favor it.

Prof. Symonds, however, came to an opposite conclusion, after a careful study of the cases that came under his observation.

The causes which predispose an animal or herd to the attacks of this disease, Collot remarks, are continued and intense cold weather, thick, damp, cold fogs, and exhalations from woods and wet places, strong currents 𝄐 air in spring and autumn, abrupt variations of temperature, exposure to rains, severe frosts, snows and storms, bad and cold, stagnant water from melted snow and ice, drunk while the animal is warm ; low, close, too warm and badly ventilated stables ; a feeding and management without change, and carried to extreme for the production of milk or labor, or insufficient nourishment followed by over-feeding, or want of regularity. Barns where the infection is known to exist ought to be cleansed in the most thorough manner, by removing all the manure, by washing with water, chloride of lime, &c., and then white-washing, and complete and long-continued ventilation for two or three months at least before it is safe to introduce healthy animals into them.

It may be proper to remark that the Dutch cattle, which seem to have been the means of introducing the disease, have suffered less severely from it than others, and the short-horns more. The Dutch is properly regarded as one of the best dairy breeds in the world ; and the fact that the disease happened to arrive with it should not prejudice the mind against it.

BLACK TONGUE.

ABOUT the time the early editions of this work were in press, another epizoötic disease broke out, and was making great havoc among the cattle of some of the southern states, especially North and South Carolina, Georgia, and Florida. In the latter state it attacked, also, and destroyed vast numbers of the deer in the forests, and was not confined to neat cattle. This malignant disease was known as the *black tongue*, and was ascribed by many to the general existence of rust in the grain and grass crops in those states. The early symptoms are stiffness, causing the animal to walk as though foundered ; copious frothing at the mouth, inability to take food, and rapid falling off in flesh, while the tongue and gums become very much swollen and turn black.

This dreadful epizoötic, unlike pleuro-pneumonia, runs its course with fearful rapidity ; and any treatment which it is proposed to try must be adopted with promptness, or it is wholly useless. It appears to be .congestive in its character, and to assume a typhoid form. As soon as the presence of the disease is suspected, Dr. Dadd rec-. ommends giving twelve ounces of table salt in one quart of warm water, adding to it two ounces of tincture of capsicum, to act as a powerful antiseptic and stimulating tonic, and to relieve the venous congestion.

Sometimes there appears to be an accumulation of gas beneath the skin. If this is observed, give the animal two ounces of pyroligneous acid, twenty-eight drops of pure oil of sassafras, and one quart of linseed tea. Mix the oil with the tea, and then add the acid. Then apply the following, rubbing the external surfaces of the tumors with it : Four ounces soft soap, half an ounce oil of sassafras dissolved in two ounces of alcohol, two ounces of tincture of capsicum, and one pint of the tincture of Peruvian bark. Cover the swollen tongue with fine salt; and as soon as any improvement in the animal's condition appears, an ounce of the fluid extract of camomile flowers may be given twice daily as a tonic to restore the appetite and the general tone of the system.

MILK SICKNESS, OR TREMBLES.

In the timbered regions of the west and in Oregon there exists a terrible disease known as *milk sickness*, or *trembles*, which disappears from the region as it becomes cleared, cultivated, and seeded down with the natural grasses. The disease is probably owing to exposure to cold, damp, and destructive exhalations from the soil, and to want of sufficient care and food — a treatment which stock is too liable to receive in the early settlement of a new country. In a section, therefore, where the disease is known to exist, the cattle ought to be housed or sheltered from the cold night air, and not turned out till the dews are dried off; and their hay or other food should not be left exposed on the ground. If after it is thus exposed to the dew it is fed to a young animal in the morning, it will be liable to cause death.

The symptoms of the disease are described as irregular nervous action, trembling, spasms, and convulsions. The pulse is quickened, the tongue slightly swollen and coated brown, the urine highly colored, the bowels constipated, and the breath fetid. In cases of constipation give ten ounces of Glauber salts, one drachm of powdered ginger, and one drachm of goldenseal, in one quart of warm water. Rub the back with a little oil of cedar. If the breath is bad, give two ounces of pyroligneous acid, four ounces of glycerine, one quart of water, mixed, a wine-glass full three or four times a day. Two drachms of tincture of Indian hemp given in a little water twice a day will relieve the trembling in cases that are curable. During this treatment the animal should be well cared for, and fed on oatmeal gruel.

Prevention is, in all cases, cheaper than cure ; and the presence of any of these epizoötic or endemic diseases ought to lead to great and constant care of stock.

INDEX.

36

www.ingramcontent.com/pod-product-compliance
Lightning Source LLC
Chambersburg PA
CBHW021347210326
41599CB00011B/784